T0291592

The Cambridge Technical Series

General Editor : P. Abbott, B.A.

A MANUAL

OF

MECHANICAL DRAWING

A MANUAL

OF

MECHANICAL DRAWING

BY

JOHN HANDSLEY DALES

Consulting Engineer; formerly Head of the Engineering
Department in Bradford Technical College ; Author of *High
Speed Engines, Principles of Steam Engines, The Superheated
Steam Locomobile, Modern Boiler Practice, Principles of
Chimneys and Chimney Draught, Transmission of Heat in
Steam Boilers,* etc.

Cambridge:
at the University Press
1914

CAMBRIDGE
UNIVERSITY PRESS

University Printing House, Cambridge CB2 8BS, United Kingdom

Cambridge University Press is part of the University of Cambridge.

It furthers the University's mission by disseminating knowledge in the pursuit of
education, learning and research at the highest international levels of excellence.

www.cambridge.org
Information on this title: www.cambridge.org/9781316606971

© Cambridge University Press 1914

First published 1914
First paperback edition 2016

A catalogue record for this publication is available from the British Library

ISBN 978-1-316-60697-1 Paperback

INTRODUCTION

MECHANICAL DRAWING is an art and craft the importance of which in the present day it would be difficult to overestimate. It would be scarcely too much to say that, without it, civilized life as it now exists would very shortly come to a standstill. In this mechanical age it has become indispensable as part of a productive system, and has a definite commercial value. It has this position, apart from its connection with Theoretical Mechanics, as a necessary feature in economic production. Indirect methods are no longer commercially admissible; every piece of work of any importance whatever begins on paper, and on the accuracy and sufficiency of what the paper shows the efficiency of the mechanism depicted there depends.

Good mechanical drawing is in itself a powerful assistance in *design*. In addition to its primary object as an accurate representation of the ideas of the designer, it presents a practical trial of the feasibility of the collective action of the parts represented. The greater part of ordinary machine design, where rigidity and practicability are the main desiderata, is produced by the showing of the drawing only, the drawing exhibiting to the judgment of the draughtsman all those elements to which mathematical considerations are not practically applicable. No complex mechanism can be designed and produced on modern lines, with complete conformability, without the aid of high grade mechanical drawing.

Scale drawing, while primarily intended for producing representations of structures in convenient sizes for drawing, handling and storing, is equally necessary to the proportionate design of all objects, which in their natural size are too large or too small to come within the field of view of the eye at the range of clear collective vision. This is especially the case where there are little or no data for mathematical calculation.

The whole of the modern necessities of mechanical production demand accurate linear drawing, displayed and explained by conventional methods, which are common knowledge to all concerned. Collectively, these constitute an important and indispensable craft, apart from any connection with abstract science, a craft which has to be acquired by labour, care and diligence. This is most readily accomplished by making it a separate study in the early stages of Engineering training; and that period is most suitable, also, because this kind of skill is a powerful assistance to the progress of every other branch of work or study which is involved in such a training.

It should also be noted that, while special aptitude is of course always a valuable asset, it is no general substitute for the necessity of studying and practising of conventional methods and expressions, and that it is therefore necessary to go through an organized training in these.

The object of this book is the systematic cultivation of high class mechanical line making, and its application as a necessary adjunct to mechanical work production by those methods and expressions which are used and understood in the drawing offices and works of Engineering firms. To this end the early part of this work is devoted to a rigid progressive disciplinary course, which by experience has been proved to be a direct means to the end of producing accurate drawing. Later, exercises in ordinary objects are introduced, and, finally, sets of drawings for complete machines provide training in drawing office practice.

The student is specially advised to spare no pains in attaining the highest possible skill in accuracy and finish at this stage, for the most important reason, that the most skilful junior is given the best work in an office, and that he has in consequence the best opportunities available for making progress and position in his profession.

An important feature of the system here adopted is that manipulative skill is acquired in a manner which admits of the concentration of the whole mental energy of the student on the acquirement of that skill, and entails no effort in understanding the figures themselves until after the time when he is well able to execute line drawing.

The artizan mechanic apprentice is advised to practise the exercises and system of this book from the beginning of his shop career, so that he will be able at the end of his first three years to change over to the drawing office should he desire to do so without any diminution, and probably some increase, of his apprentice wages. By reasonable industry he will thus be able to widen the outlook of his future career as well as to lay a solid foundation for high scientific attainments, and eventually be able to reap the full benefits of his work.

J. H. D.

June, 1914.

CONTENTS

CHAPTER I

CHAPTER II

CHAPTER III

CHAPTER IV

CHAPTER V

CHAPTER VI

EXERCISES

Contents ix

CHAPTER VII

OBJECT DRAWING

CHAPTER VIII

WORKING DRAWINGS

x *Contents*

LIST OF ILLUSTRATIONS

MECHANICAL DRAWING

CHAPTER I

The scope of mechanical drawing is not necessarily confined to the delineation of machinery, as it is both useful and necessary in any branch of constructional art where accuracy is required.

The principal aim of the student of this branch of engineering craft should be absolute accuracy in position of his lines, and clear and full description by dimensions and notes of what the lines represent. Nothing at all should be left to the imagination of the workman, whose business is solely to carry out the shewing of the drawing in the materials of construction. The drawing should be measurable at every point by a scale corresponding, and the dimensioning should correspond to the drawing both locally and collectively; the dimensioning should be regarded more as a convenience in reading off sizes and distances than as any substitute for inaccuracies of drawing. The drawing should be a check on the dimensions, and the dimensions a check on the drawing.

For the attainment of that manipulative exactness which is needed, all the apparatus and implements

which are used should be of the best quality obtainable and of such patterns and make as will make the
production of exact work a possibility, with ordinary
care and commercial expedition.　Patterns which are
difficult to handle, or which require both hands instead
of one for adjustment in working, or in any way are less
convenient than such things can be made, are quite in-
admissible and are not in any circumstances to be used,
as they are eventually a handicap on the draughtsman's
means of getting a living, and are dear at any price.
Inferior implements tend to produce bad habits in manipulation which are difficult to get rid of when once acquired.
The apparatus and instruments are the tools of his craft,
and the cost of the best is a negligible item in the value
of the work which is produced by their use.　No man in
competition can afford to work under any heavier handicap
than his competitor.

All mechanical drawings, or nearly all, contain work
of a great variety of difficulty, some parts being relatively
easy to the beginner, and other parts of a degree of
difficulty sufficient to tax the skill of an accomplished
draughtsman.　This variety has always prevented the
student from making any satisfactory progress in his
work of copying drawings of mechanical objects for practice for a considerable length of time, as nothing collectively passable could be done until the more difficult
minutiæ were mastered; and these occurring only seldom
in the figures, and therefore affording relatively little
practice, a very large amount of waste labour has had
to be expended in repetitions under the old system of
teaching before sufficient proficiency for practical purposes
could be attained.　In order to get over these difficulties
and expedite the rate of progress in the early stages of

study, the author some years ago devised a series of
progressive exercises in the use of scales and instruments
which have invariably shortened the time for acquiring
manipulative skill, from twelve months to three months,
and in cases of exceptional aptitude to even much less
than the latter time. This is a very important matter in
many ways. The saving of time is great, and especially
valuable at a student's time of life; but the saving of
patience and avoiding of vexatious and repeated failures
to execute the more difficult parts of an ordinary object
drawing, and thus repeatedly wasting the whole, are
perhaps even more important. Many students under the
old system have given up the attempt to become good
draughtsmen, and many more have lapsed into loose ways
of finish which have never left them.

LINE EXERCISES.

The Exercises, which form part of this work, are
arranged to give a large amount of practice in each stage
on small sheets of paper, so that those containing irremedi-
able "slips" can be destroyed and a fresh start made
without much loss of either time, patience, or material.
They develop a sufficient skill for object drawing before
such work is attempted, and at the same time a skill
in the minutiæ which is altogether superior to that which
is usually acquired under the old system. Also these
exercises entail no mental effort in the understanding of
the figures, and so leave the whole mental energy available for concentration on the manipulation. This is
important, and adds to the rate of progress in skill. The
repetitions also provide opportunity for the student to "ex-
periment" with the instruments, and so to find out what
they will do and what they will not do. The teacher's

1--2

work in the opening stages of this study is very much less than under the old system, as the simple use of the scale and an examination of the quality of the lines are all that it is necessary for him to do while work on the exercises is going on, and he can consequently give efficient supervision to larger classes.

CHAPTER II

APPARATUS AND MATERIALS FOR DRAWING

DRAWING BOARD.

Among the apparatus and instruments which are employed in mechanical drawing, the drawing-board is the first to be considered. This is a perfectly rectangular board, made of the best yellow pine, about 1 inch in thickness. It is grooved at the back and stayed with oak or mahogany cross-pieces or clamps for preventing skellering or twisting from damp and changes of temperature, which would spoil the flatness of the working side. The sizes are such as are required for the paper drawn upon; but in drawing offices the most common size in use is that which corresponds with the "double elephant" size of paper, which is 40 inches long by 27 inches wide, the boards being made 42 inches by 28 inches. This size is as large as the draughtsman can reach over comfortably, but in some offices 60 by 30-inch boards are used. For special purposes some are even larger.

Wherever there is room for its use, it is advisable that all drawing practice should be done on the regular

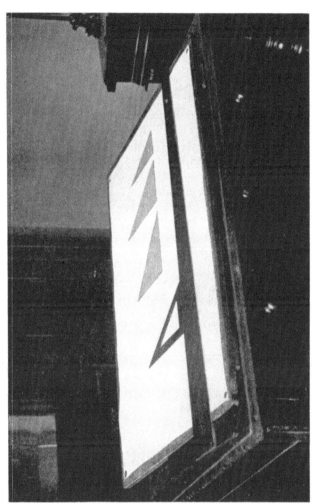

Fig. 1. Drawing-board, T-square and Set-squares.

drawing office size of board, in order that the pupil may become accustomed to handling the ordinary commercial apparatus. The drawing paper is fixed on the drawing-board by means of short pins with large heads, the pins being pressed through the paper at the corners into the soft wood of the board. Formerly drawing paper was damped at the back, which expanded it, and then it was glued to the board all round its edges When dried the contraction stretched the paper very tight and straight. But in those days more time was spent on drawing, and often artistic work and colouring was indulged in, while in the present utility and expedition are the desiderata only. It is very seldom that any costs are incurred which can be avoided. Moreover, elaborate mechanical drawings are sufficiently expensive even when they are produced in the cheapest possible way.

A drawing-board as described is shewn by Fig. 1.

T-Squares and Set-Squares.

The next, and very important item in the equipment, is the T-square. This is a perfect straight-edge, having a T end which slides against the edge of the board when in use. The T end being held in close contact with the edge of the board, maintains the drawing edge of the T-square at right angles to the edge of the board, and the draughtsman is thus enabled to draw across the drawing paper, lines of any length which are square to the edge of the board and perfectly parallel to each other For drawing rectangular figures it only remains to provide a similar method of making lines "square" to the T-square edge as the T-square lines are to the edge of the board. This is accomplished by the use of a "set-square," which is usually a triangular sheet of thin material (wood or vulcanite), which has two sides at right angles, the angles being usually 45°, 30° and 60°, or $22\frac{1}{2}$° and $67\frac{1}{2}$°, each pair of angles, of course, making 90°. The T-square and set-squares (of which latter it is usual to have a set) are shown in position on the board in Fig. 1.

The T-square and set-squares are manipulated by the left hand of the draughtsman, the friction of the fingers maintaining a slight pull on the blade or straight-edge for the purpose of preserving close contact of the "butt" or T end with the edge of the board, while the set-square is placed in the position which the work requires by the right hand, and is held there by the forefinger of the left hand. The left hand is thus fully occupied, while the right hand is free to use the pencil. Horizontal lines or lines across the board are usually drawn from left to right, but it is always necessary to *see* the point of the pencil at the finish or "joining up" of a line, and the end of a line is often drawn back from right to left and "joined up" so that it may not be necessary to rub out "over-drawn" lines—lines which pass the "set-out" marks. In some cases it may be convenient to "over-draw" one or both sets of lines (vertical and horizontal) very lightly, and then go over the "correct figure" portions with a heavier line. The fine "over-drawn" lines are a guide to the exact finishing points of the ink lines, where the finishing points are difficult to see by reason of the hand or ruling edges coming into the line of sight.

The holding of the T-square and set-square is an important matter, and is shewn by Fig. 2. It is usual and the regulation method for the student to learn to "rule," or draw the lines, from bottom to top—to rule *away* from the T-square when using the set-square. But this is not imperative, though generally convenient and

Fig. 2.   Holding T-square and Set-square.

less liable to "joggle" the elbow of the draughtsman, and so spoil the line, especially when "inking in" with the ruling pen. When the pencil is used with the right hand—which is most usual—and the set-square lines are "drawn away" as stated, the set-square is laid with the ruling edge to the left hand, as shown in Fig. 2. Sometimes, however, it is necessary or expedient to draw lines "down" with the set-square, and in such case it is usually laid with the ruling edge to the right. A due observance of such minutiæ at the outset greatly assists in the cultivation of unconscious habits of manipulation which are so necessary in any kind of craft.

Both T-square and set-squares should be made of the best and well seasoned material, in order that they may not be affected by damp and change of temperature, and in use they should lie perfectly flat and in close contact with the paper, with very little pressure from the hand of the draughtsman. They should, of course, be perfectly straight on the edges and exactly true in the angles, the quality of the drawing depending of course on this. The T-square should be edged with ebony and the body made of mahogany, and large set-squares of the same materials, while the smaller set-squares are now made of clear celluloid or black vulcanite. The black vulcanite kind "dry in" less and keep true the greatest length of time, but the celluloid are most convenient to use, as the lines of the drawing can be seen through the material, and thus accidental "over-drawing" is often avoided. All the "ruling edges" should be about $\frac{1}{16}$ of an inch deep, neither more nor less, and both T-square and set-square edges should be the same. This is a point which materially assists the draughtsman in expeditious and correct working, for, as the pencil point and the pen point have each some thickness, they cannot be laid dead on to the line of the ruler edge, and some practice is required in judging the distance of the marking out from the edge, so that the lines may be made in correct position without alteration. The depth of the " edge " is also required for occasional " humouring " of the joining up of lines by the "in" or "out" inclination of the pencil point, and in the case of inking, the pen can be inclined "in" or "out" to bring both nibs on to the paper and so ensure a solid line. If the edges are too thin the adjustment of the distance of the line is unduly difficult and leads to slower work, while if they are too thick they are equally inconvenient by making the proper distance from the marking out too indefinite and difficult to judge. Stress has been laid on the fact that this kind of drawing is a craft with a market value; hence convenience or inconvenience in use of the apparatus directly affects the value of the draughtsman's " time."

Common school T-squares and set-squares are made of pear-tree wood, but such things are of no use to the professional draughtsman who is in competition with other men who are well provided; and the best quality of apparatus lasts practically a lifetime.

Clean Drawings.

As most mechanical drawings require considerable amounts of time for their execution, and consequently keeping such drawings clean is something of a problem, the paper should be well dusted and rubbed with a clean duster each time when work is commenced, and both the T-square and set-squares should be frequently cleaned and kept quite clean, in order that " blacks " and dust may not be rubbed into the drawing. This cleanliness is not easy to maintain, but it is well worth the trouble, and is actually economical in time, as on a "long job" it is often necessary otherwise to "clean a place," so that the draughtsman can see what he is doing. The rubbing required in so doing spoils the surface of the paper and makes a good finish difficult or impracticable, while in tracing from pencil drawing, dirt makes it difficult to see the lines. Care should be taken to preserve the surface of the paper by using only soft india-rubber in a regular way; the harder or gritty erasers should be used only for taking out very heavy or ink lines. It is more economical in time to be careful to avoid mistakes than to have to correct such as are made by undue hurry. But herein there is a common-sense medium; rules are general, not arbitrary.

PAPER.

There can be no doubt that the best quality of paper obtainable is the most economical in the ultimate cost of a drawing. Good, accurate work is imperative, and this can be produced most expeditiously on good paper. The best paper for this purpose costs about 9s. 6d. per quire, double elephant size. The cheapest quality will cost about half that amount. The difference is 4s. 9d. in 24 sheets, or about $2\frac{3}{8}d$. per sheet. The cash value of a double elephant size drawing may be anything from 20s. to £20, or even more; so that the cost of the paper is a negligible item, and good against bad paper may easily make a difference of 20 per cent. in the time required for a drawing. If any "accident" happens and anything has to be erased, the bad paper is hopeless, while a passable job may be made on good paper.

There are many excellent papers on the market; and strongly sized (stiffened) papers of good strength and

substance are to be preferred. For plain line drawing—
no colouring—smooth, hard surfaces are the best. Thin
or soft paper is to be avoided, as the compass-points go
through too readily, and the hard, sharp-pointed pencils
which have to be used also cut indelible marks in the
surface. Poor paper is thus much dearer in the end.

The standard "surfaces" of drawing papers are such
as correspond to Whatman's "hot-pressed," "not," and
"rough." The two first, "hot-pressed" and "not," are
those which are used for mechanical drawing, and the
student is advised to have "Whatman" paper, at least
until such time as he comes to know the different qualities
of the different "makes." On the hardest and smoothest
papers, pencils of moderate hardness, say 3 H, can be used
with advantage, as the lines are blacker and easier to see
than those made with 4 H, which are used on rougher
paper. A little experience, however, soon shews the student
what hardness of pencil best suits his hand. Rough surface paper takes colour better than smooth paper, especially
if the latter has been rubbed, when the colour runs
"cloudy." Rough paper is therefore sometimes used for
"show" drawings.

The draughtsman is advised to employ as soft a pencil
as is practicable with the paper which he uses, as fine line
drawing is trying to the eyes in any case, and the soft
pencil makes a blacker and more distinct line than the
harder variety. The term "soft" is applied to 3 H of the
best makes.

CHAPTER III

SCALES AND SCALING

As it is very seldom that mechanical objects can be drawn full or natural size, it is generally necessary to draw to some fraction of that size, which is technically termed the "scale" of the drawing. This scale is made to bring the object on to a convenient size of paper, or of such a size that detail can be plainly shewn, and dimensions and references can be got into the spaces to which they refer. For mechanical work, marked straight-edges made of boxwood or ivory are used, and these are termed *scales*. The markings are in some proportion to the markings of a 1-foot rule, full size. Thus, some scales are set out with principal divisions of 3 inches each, and a 12-inch long scale (commonly termed a 12-inch scale) would contain four 3-inch spaces, each divided into twelfths to represent inches, which are sub-divided into halves, quarters and eighths, in the same way as an ordinary full-size rule. The 12-inch, 3-inch scale would thus span 4 feet to 3-inch scale. In like manner scales are made with 6 inches representing a foot, generally termed "half-size," the 3 inch before mentioned, $1\frac{1}{2}$ inches, $\frac{3}{4}$ inch, and $\frac{3}{8}$ inch. The latter is generally considered to be as small as is practicable, as the 1-inch divisions are actually only $\frac{1}{32}$ of an inch, which could not ordinarily be conveniently sub-divided to be of any use in regular work. As these proportions are multiples of eighths, being $\frac{1}{2}$, $\frac{1}{4}$, $\frac{1}{8}$, $\frac{1}{16}$, and $\frac{1}{32}$ of full size, they are termed *eighth* scales. In like manner scales are made of 2 inches, 1 inch, $\frac{1}{2}$ inch, $\frac{1}{4}$ inch, and $\frac{1}{8}$ inch to a foot, which, being multiples of twelfths, are termed *twelfth* scales. These are two of the dividings generally used in British drawing offices for machine draughting. But occasionally when work for export is being done, metric scales are used, which are made to some proportion of a metre, in a similar manner to that which has been described in reference to British scales. For special purposes, decimal scales to a foot or inch (generally the latter, especially in America) are used. Special scales may of course be multiplied indefinitely, but are not commonly procurable out of stock and have to be specially made. Chain scales and the like are used in Civil Engineering. In British practice, however, the eighth and twelfth scales are those generally used, and they are made commonly 18 inches, 12 inches, and 6 inches long, and occasionally 2 feet and 3 feet long. Speaking generally, $\frac{1}{64}$ of an inch actually is the smallest division which can be " set out " on a drawing from a Scale, to be of any practical use; and it is seldom that divisions are less than $\frac{1}{32}$ of an inch, these, where required, being divided on a line, or from a point, by sight. Thus $3\frac{1}{2}$ inches to $\frac{3}{8}$ inch scale would be $\frac{7}{64}$ inch actual, and at any point from which such a distance had to be set out, the Scale would be laid so as to divide 1 inch, or $\frac{1}{32}$ inch actual, on the point, the " set out " mark coming then on a $\frac{1}{32}$nd (or 1 inch) division.

Scales are made fully divided or open divided. Fully divided scales are marked with the smallest divisions for the full length of the scale, so that from each end any measurement can be read off anywhere, while *open* divided scales are divided to the smallest division only at each end of the scale, that is, only 1 foot (to scale) is fully divided. .The remainder of the length of the scale is marked or divided only at such distances as represent

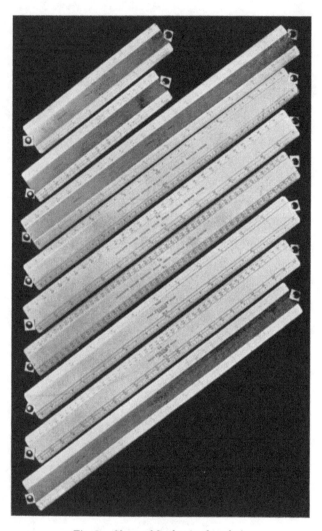

Fig. 3.   Sheet of Scales (scale rules).

1 foot. Thus a 12-inch, 1-inch scale, would at each end be divided for the length of 1 inch to, say, $\frac{1}{4}$ inch (48ths actually), and the remainder of the intervening space would be divided into 1-inch spaces, each representing 1 foot. There would thus be twelve spaces, representing 12 feet, with only the two end spaces fully divided, and in measuring, say, 9 feet $8\frac{1}{2}$ inches to scale, the ninth open space would be laid on the off-set mark and the $8\frac{1}{2}$ inches would be read off, or marked off, on one of the end spaces which are fully divided. Fig. 3 shews a number of scales such as are mentioned, the principles of which will be readily understood, and by inference, the construction of any other.

The best material for scales is undoubtedly ivory, but this is expensive. If it can be afforded it is the cheapest in the long run. Ivory scales cost about three times as much as boxwood, or boxwood edged with celluloid. The divisions on scales should be cut well into the material, and the edges of the scales should be made as thin as practicable. The "oval" section is the best for working. The dividing on wooden scales should be done after the varnishing, so as to leave the divisions clear for taking the sharp edge of the pencil in setting out, of which more will be said later.

As to the "dividing" on scales, it is inadvisable to have more than one kind of divisions, that is to say, more than one kind of the smallest division on each scale or rule. For instance, the "eighth" scales are ordinarily made up of 32nds of an inch as the smallest divisions. At $\frac{3}{8}$ inch to a foot $\frac{1}{32}$ represents 1 inch, at $\frac{3}{4}$ inch to a foot $\frac{1}{32}$ is half an inch, at $1\frac{1}{2}$ inches to a foot $\frac{1}{32}$ is a quarter of an inch, and so on. A scale or rule may have 3 inches to a foot on one edge of a side, $1\frac{1}{2}$ inches on the opposite edge of the same side, $\frac{3}{4}$ inch on one edge of the reverse side, and $\frac{3}{8}$ inch on the other edge. All the small divisions would be $\frac{1}{32}$ of an inch, so that in the abstraction of his work the draughtsman is not liable to set out wrong distances, as he is when the scale or rule has " eighth " divisions on one side and " twelfth " divisions on the other. Constant turning over of the rule and " sorting out " of the right division is a great inconvenience and nuisance to the draughtsman when his mind is concentrated on the work in hand. Two kinds of scale rules may therefore be said to be necessary in the equipment of a draughtsman, one with " eighth " divisions and one with " twelfths," and also two lengths of each kind, 6 inches and 12 inches, as the longer length is often unwieldy for close or small work. A long scale 18 inches or 24 inches is often very useful for over-all dimensions.

CHAPTER IV

DRAWING PENCILS

The next important implement to be considered is the drawing pencil—a most important implement.

For actual mechanical drawing no note need be taken of the softer varieties which are used for " art " purposes, nor of those which are used for figuring or writing, as practically anything which will take a point and keep it reasonably will do for such matters. The *draughtsman's* pencil is a *tool*, a fact which must be insisted upon and kept in view. Good lines are clear and distinct and exceedingly fine, as made by a good, properly sharpened

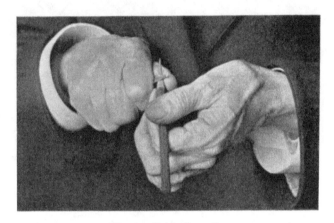

Fig. 4.   Cutting pencil.

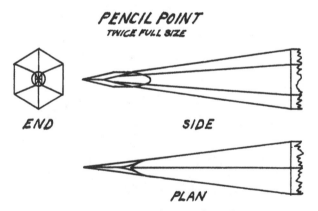

PENCIL POINT
TWICE FULL SIZE

END   SIDE

PLAN

Fig. 5.   Enlarged point of pencil.

pencil. They are finer than any inking which can be
done, and finer and more correct in position than any
practical work can be executed—note, "correct in position."
Admittedly, very close jointing can be done by good fitting
or grinding, but taking a number of joints and positions
at distances, no commercial practical work is ever so
accurate as a good drawing when first executed. There
is no desire to split hairs on this point; what is aimed at
by the statement is to impress the mind of the student
with the importance of the pencil as a *tool*. What the
pencil does, or fails to do, cannot practically be improved
by any subsequent operation.

As the cost of the pencil is no appreciable item, and
as it is a time-saving or time-wasting *tool*, the absolute
best which can be obtained is the only kind worth considering. There are not many "makes" which are of
first-class quality; and as it is necessary at the outset to
direct the student's selection, the harder grades of Wolff's
and Hardtmuth's manufacture may be mentioned as in
general use. They are both stated to be made of British
graphite, and are indeed excellent productions. The
" lead " should be smooth and free from grit, and black—
not grey; it should be of a tough texture, sufficient to
stand working when cut with a very fine long point.

The cutting of the pencil point is an important
operation requiring skill and care, and which can only
be performed properly with a knife made with a "chisel "
edge properly "set " or sharpened. A knife-cut point
makes a "cleaner" line and lasts very much longer than
a rough-cut point rubbed up on a file or glass-paper.
The lasting properties of a pencil point are, in the matter
of time saving, the most important, while the fineness and
clearness of the lines made by it are scarcely less so. A

proper method of holding the pencil when cutting the point is shewn by Fig. 4, and a correct form of point by Fig. 5, which is drawn twice full size for the sake of clearness. In this matter, as in all others connected with the apparatus, it is important to avoid any defects in the tools which may divert the draughtsman's mind and prevent full concentration on the work which he has in hand.

The pencil point, including the wood, should be 1⅛ inches long, and cut slightly "hollow," or curved, so as to expose ⅜ inch of lead, which on two opposite sides should be *cut*, first to produce a "chisel" point the width of the lead. The sides of the point should then be cut to form sharp edges, the general form of the point then being that of a double-edged dagger, with nearly straight edges, sharp on the edges and at the point. The edges are cut for the purpose of engaging in the dividing lines of the scale when marking off distances on the drawing, such lines, as before stated, being cut well into the material of the scale for the purpose of guiding the pencil edges.

In using the pencil for "marking off" distances on the drawing, the scale is tilted so that its edge lies close on the paper, near the line on which the marks are to be made, and the sharp *edge* of the pencil is then "run down" the division lines of the scale on to the paper, thus marking the distances desired with maximum expedition and practically dead accuracy (see Fig. 6). The pencil must be held firmly in the fingers and in one position throughout any "marking off," so that any deviation from squareness or uprightness of the edge of the pencil may not interfere with the relative accuracy of the markings. The knife used for cutting should be "dead sharp," and the cutting edge of "chisel" form, that is, flat on the side used

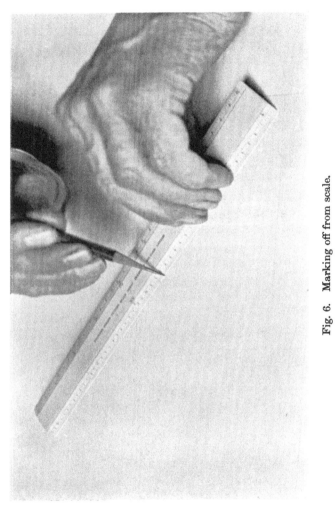

Fig. 6. Marking off from scale.

next to the pencil and with the edge angle on the outer side. This is very important and an essential to correct and expeditious cutting. The wood and lead can be cut "right through" with this form of knife, and the point can be very quickly prepared.

Fig. 2 (Chap. II) shews the correct method of holding the pencil when working with it, either with the T-square or set-square. It will be noticed that the little finger rests upon the ruling edge and supports the weight of the hand as well as acting as a gauge to the pressure put upon the pencil point—very light for a fine line and heavier for a strong line. Also it is used for keeping the ruling edge close down on the paper and the paper solid on the board, which is necessary for ensuring correct work, as the ruling edge is seldom dead flat, nor the paper without some "buckle." The pencil can be used very "close in" to the ruling edge—much closer than a pen. The pencil must be kept quite sharp by frequent scraping with the edge of the knife, and especially so when used for marking off the scale.

All the pencil lines should be made as they are intended to be inked in, that is, the drawing should be as well "finished" in pencil as when inked, and especially when tracing has to be done from the pencil drawing, which in the present day is almost the universal rule, the tracings being made on tracing cloth for subsequent blue printing. These remarks on "finish" in pencil drawing refer to dotted work as well as to any other part of a drawing. As a matter of fact, good "close work" (crowded lines) cannot either be executed or traced without clear finish in the pencil lines. Good pencil work is much more economical in cost eventually than inferior work, though it might be supposed that the latter would take less time; but it does not, the better work assisting its own progress.

Drawing Instruments.

Under this heading are included mainly such implements as are used for making circular lines, circles, and
curves. Mostly they consist of what are commonly termed
compasses. Typically they are jointed legs, with centre
point on the end of one leg and a drawing point for pencil
or ink on the other. Fig. 7 shews an ordinary "set" of
compasses. The large compasses (1) are termed "dividers,"
having both legs fitted with centre points, and are used for
dividing lines or distances which cannot conveniently be
set out by the scale. (2) is a similar instrument, but is
made with knee joints and changeable ends to one leg, for
either pencil or pen, and for inserting a lengthening piece
for increasing the range of the compasses. This instrument (2) and its parts are termed a "half-set," 5 or 6
inches long in the legs. (3) and (4) are "bow" compasses,
being similar to the foregoing in general form, but made
with a fixed permanent end on one leg for pencil or ink,
as the case may be, the pencil fitted being termed the
"pencil" bow and the pen fitted, the "ink" bow. These
are made in different lengths from 3 to $3\frac{3}{4}$ inches. The
longer sizes are fitted with knee joints to the legs (see 3a
and 4a), and any "size" is better for having knees, as the
centre point can then be kept more at right angles to
the paper, and thus does not make a large centre hole,
while the pen point can be adjusted to bring both nibs on
to the paper and so make a solid line. For making very
small curves or circles, or a number of exactly the same
size with certainty, smaller compasses with spring legs are
used, (5), (6), (7), which are termed spring bows and
dividers. These have no knees and can be used only to
a limited range of radius. (8) is a separate pen for ruling

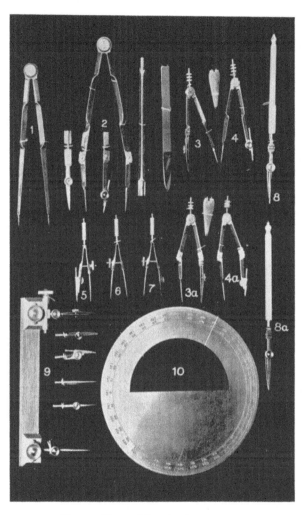

Fig. 7. Sheet of drawing instruments.

lines with the T-square or set-square, or with set curves, that is, curved rulers. (9) shews a pair of clamps which are provided with pencil and pen fittings, and which can be fixed on a rod or "beam" for making large curves or circles. These are termed "beam compasses," and are made with various modifications. For long radius circles "beam compasses" are to be much preferred to the "half-set" used with the "lengthening bar," but they are not very portable, and are usually "office tools"—belong to the office for common use. (10) is a protractor for setting out angles.

In providing himself with instruments the student should not select any but the *best quality*, as with none other can he hope to attain to that high skill and expedition in work which is necessary to pass muster in an ordinary professional drawing office. The whole expense of a set of instruments, scales, and set-squares, is not great, considered as the means of doing work which has a relatively high commercial value, and it is mistaken economy to have tools which are a handicap in what is more or less commercial competition.

The young beginner, with whom the cost of a whole set may be a consideration, can get through the "learning" stage and become an expert line-maker with very few instruments. A T-square, 60°, 67½°, and 45°, 6-inch set-squares, one each pencil and ink, 3¾-inch best bows with double knees and solid points, and two 6-inch scales of eighths and twelfths as described, are really all which are absolutely necessary, a ruling pen being easily extemporized by "bending in" the "centre" leg of the "bow pen." Good instruments of this size can easily be set to make the smallest circles. The larger circles can be made by the "school" beam compasses, which are usually provided.

All these instruments, etc., are suitable for "office" work, and can be added to as circumstances permit. In no case is it any practical use to attempt the attainment of high-class work at a commercial rate, with bad tools. This insistence on good instruments is as necessary as any other essential discipline in a training course. No man can reasonably hope to become a workman good enough to earn a standard wage if he has bad tools with which to learn his business.

The "setting" of the instruments is an important matter, and the pupil should learn to do this himself, so that he can constantly keep his tools in proper working condition. A very fine emery "slip" should be kept for the purpose as part of the equipment. The "centre" leg point should be kept always "dead sharp," the actual point invisible to the naked eye, so that it will "hold" in the paper, sufficiently for making a curve or circle without making any larger hole than can be easily distinguished at close quarters. In "trying" for a centre, care should be taken to avoid making a hole, and when the centre is found, the impression of the compass point should be just sufficient to enable it to be found when the inking is done. The "tools" should be lightly but firmly handled, the little finger of the right hand resting on the paper for regulating the pressure and guiding the points to the "marking out." When several circles are made with one centre great care should be exercised not to press the centre point into the paper any more than it is necessary for preserving the position of the centre point, and the adjustment of the pencil or inking leg to the radius required should be made closely approximate before inserting the point in the centre hole. If the adjustment is done with the leg in the centre hole the hole will almost certainly be enlarged to the detriment of the drawing. Any slight adjustment necessary should be very carefully made as regards pressure at the side of the hole.

In the question of solid *versus* needle points for the compass legs, the writer unhesitatingly pronounces for the former, which he has always used both in his own work and in teaching. Solid points are certain and definite in action *always*. Needle points may or may not be, as likely one as the other; and as facility for renewing the length of the "centre" leg is the only object, and an unnecessary object, there is no advantage to be gained by having a "loose end" which may give out at a critical moment; one which is always more or less "lumpy" near the point and prevents very close placing of the pencil or inking leg for making very small centres. The idea that a needle makes a hole which gives a more correct action of the working leg than a solid end is erroneous. The contrary is generally the case. A needle can scarcely be prevented from working too deeply or *through* the paper, and in the latter case the point wobbles around underneath and makes a hole which has no actual holding centre. The needle is actually too fine in the shank and goes in too deeply. The solid point, on the other hand, if rightly set, is conical, and fills the hole as it sinks, thus remaining concentric with the lines which are drawn. As to the needle affording facilities for renewal, no renewal is required. The pencil point is always adjusted to the centre point, whether solid or needle, and the pen point always wears down more rapidly than the centre point, and therefore there is always a margin for resetting the centre point; so the needle in that case has no advantage. With reasonable care, which is always necessary for preserving the working condition of the instrument, there is practically no risk of the solid point getting damaged beyond the possibility of resetting. In placing the centre point on a mark or intersection it is necessary that the point should be clearly visible. and a "shiny" needle rather defeats than assists that delicate operation.

Anything connected with the working efficiency of the tools has a direct bearing on the expedition of a draughtsman's work, and also on the quality both of the drawing and the embodiment of principles of the matter in hand, so that even small distractions may lead to much general inconvenience. Therefore, what may appear on the surface to be trivial matters are actually important.

MANIPULATION.

A proper method of holding an instrument when setting the centre leg or point on an intersection is shewn by Fig. 8, in which the little finger rests upon the paper and acts as a guide to placing the centre point correctly. This method applies to the placing of all the instruments which are used for drawing circular lines.

The inking of a drawing should be solid and black, which can only be ensured by a proper consistency of the ink. Rubbed ink is much the best for securing this result. If the ink is too thin, the lines, even if solid, will be grey, and if too thick will not run either in fine lines or sufficiently free for expeditious work. A good guide for judgment of the proper consistency is that when just thick enough to be black, the ink cannot be blown by the breath off the surface of the inkpot or dish. Liquid Indian ink is seldom sufficiently thick to be quite black, and the lines are liable to be rubbed off to greyness when the drawing is cleaned off. Especially is this blackness and density of

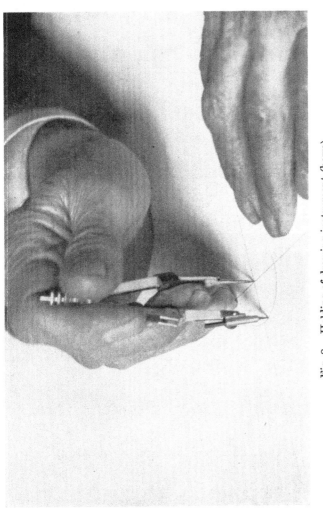

Fig. 8. Holding of drawing instrument (bows).

the lines important in tracings for sun printing. Sometimes a little yellow colour is rubbed in with the ink
which is to be used for this purpose in order that any
light which penetrates may be non-actinic. Lines which
are to be made to coloured sections may be prevented
from "running" by adding a small quantity of alum to
the ink or to the colour. The addition of colour to the
ink may cause it to "separate" and become both watery
and lumpy. Very fine lining cannot be done with a thick-
nibbed pen or with a pen in which the nibs are too
"parallel." The pen should be "bellied" sufficiently to
allow the ink to run down to the point freely when the
nib points are very close; otherwise capillary attraction
will prevent the ink from either running to a fine line at
all or at a sufficient rate for expeditious working. In the
drawing of whole circles the line can seldom be made at
one circular sweep of the compass, either when solid or
dotted; and the finish or joining must always be in clear
sight. It is therefore generally best to make the first
part of the sweep "over" clock wise, about two-thirds of
the whole circle, and then to join up the gap by a backward
stroke.

Dotted lines as commonly understood are not really
dotted (which is too slow a process) but dashed. The
length of the dashes are according to the size of the
portion of work in which they occur, but in any case, to
ensure neatness and visual continuity, the spaces between
the dashes must be very small—just clearly visible.

In the actual drawing of the lines with the T-square,
either by the pencil or inking pen, it is expedient to make
a trial of the distance of the "setting out" in the pencil
case, by moving the ruling edge until the proper inclination
of the pencil "over" the edge is obtained, rather than to

endeavour to adjust the inclination of the pencil to the position into which the ruling edge happens to fall at the first trial. The pencil point should be as close "in" to the edge as possible, since in that position it is least liable to make the line out of parallelism as the line proceeds, especially if it is a long one, by any alteration in the inclination "in" or "out" of the pencil. This is particularly important in putting in centre lines which determine the positions of the rest. This note is made to counteract any erroneous supposition that in drawing on paper with a T-square the pencil point can be placed on the "mark" and the T-square then brought up to it. A pencil point of the proper sharpness is much too delicate to stand the shock of arresting the motion of an article of the weight of the T-square, when that is brought up with sufficient quickness for practical work. The point would be broken or moved out of place by any attempt to use it in such a manner. In like manner the pen would be moved out of place. In ruling lines with the "set-square," which is relatively very light in weight and also more difficult by reason of its position, and even lightness, to place in exact position to suit the thickness of the pencil or pen point, the points of either are best placed at the proper distance from the actual edge of the ruler by setting them *on* the mark, and then bringing up the "set-square" with the forefinger of the left hand, as a final adjustment, into contact with the pencil point for making the lines.

For drawing circles of an exact given diameter, the *radius* is first taken direct from the scale by the compasses. The resulting *diameter* should then be *tried* on the centre line (if the circle is large) or by making a circle with the "setting" of the compasses (if it is small) and then checking by the scale right across the diameter. Any error of the "setting" of the compasses will of course be doubled on the diameter, and when exactness is imperative it must be got by adjustment before the circular line is made on the drawing. By quick movements these precautions can be taken in less time than alterations can be made to a wrong diameter, the result being correct and certain.

With reference to exact instructions as to method, etc., the student should realize that they are framed for the purpose of cultivating a *professional skill* as expeditiously as possible. When he has attained to a professional skill, of course he will use it according to his judgment, either fully or partially, according to the importance of the work which he happens to be doing. Instructions are also framed for preventing the acquirement of bungling methods which never lead to high professional quality of work, but which waste time and develop bad habits of manipulation.

Before commencing to work on a drawing the draughtsman should clean and prepare all the apparatus and instruments which he is about to use. The preparation of the pencil point and the adjustment of the compass pencils to the length of the centre points are the main items. The centre points should be longer than the pencil point by about $\frac{1}{100}$th of an inch, so that very small circles can be readily made without any special setting at the moment of use. The pencil point should be cut in the same manner as described for the ruling pencil, and should be carefully set "square" to the path of the line drawn, so as not to have any liability to run "in" or "out" of truth. A properly sharpened and adjusted pencil compass will make a clean circle $\frac{1}{50}$th of an inch in diameter without any trouble. Ordinary Whatman paper is about $\frac{7}{1000}$ths

of an inch in thickness, so that $\frac{1}{100}$th of an inch projection
of the point would go through. Nothing should be left
unprepared which will hinder the progress of the draughtsman's work or divert his mind from the work in hand, as
he has to think about, and have a reason for, every line
which he produces, and the smallest distraction may lead
to serious consequences.

A draughtsman's work proper is the transference of
ideas to paper, and his manipulative habits and skill
should be such that they are exercised unconsciously;
hence the necessity for systematic methodical training.

STUDENTS' DRAWING INSTRUMENTS.

The provision of drawing instruments, which are at the
same time of first-class working quality and of a price
which comes within the means of the average evening class
student or artizan apprentice, has always been a grave difficulty with technical teachers, and many attempts have been
made to provide a combined instrument or set of instruments which would be sufficiently efficient for the early
stages of study in this subject. The combined instrument
was usually a small half set. This however has never been
fully satisfactory. The principal reason for this has been
that the 3¾ "bow," of Stanley's original pattern, is in itself
a perfect instrument for its purpose, and therefore cannot
be made either larger or smaller without detracting
seriously from the features of efficiency which the double-
kneed type possesses. If it is made larger it cannot be
handled with the exactness and expedition which professional draughtsmanship demands—and imperatively
demands; nor at its proper size for handling, is there
any sufficient length for the necessary features and fittings
for loose parts. The "thumb gap" in the upper part of

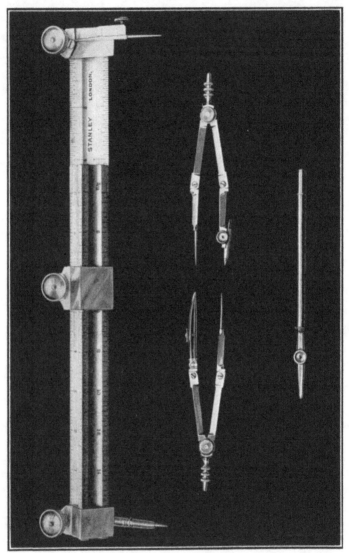

Fig. 9. Students' Drawing Instruments.

the legs of the $3\frac{3}{4}$ "bow" is not any too long, and in the lower part below the joint there is only just sufficient length for a "pen" of the proper and necessary size. A longer instrument cannot be held properly by the milled top, and at the same time the paper be reached by the little finger for guidance of the centre point. The handling of such an instrument entails placing the point with a "short hold" by the centre leg, and then an extra movement of the hand in taking a hold on the milled head, which takes up more time and causes a "heavier handling" of the instrument. The author has had many such instruments made, in endeavours to produce something which would do both large and small work, but nothing of a really practical character has so far been evolved. It has become evident that the Stanley pattern instrument (which has now been in use universally for over 50 years) cannot be either altered or improved upon, and that, within its range, it must remain as it is. The minimum requirement therefore of the student, is two "bows" of this description; and for larger work, it has become evident that something cheap and firm in the way of a beam compass is the next minimum alternative. To provide this latter, the author has suggested to the firm of W. H. Stanley & Co., Ltd., that they should produce a pair of beam compass heads which will clamp on to an ordinary triangular scale, or on to a triangular wood rod which can be easily procured; all at an absolute minimum price for the necessary quality required. They have done this, and have produced a set of instruments with which the whole of the work in this book can be done. As a matter of fact the whole of the drawings for the blocks herein contained have been prepared with instruments such as are shewn in Fig. 9.

DESCRIPTION OF FIG. 9.

This shews two triangular scales marked on 6 in. of their length, with a blank at each end of about ⅝ in. The beam compass heads can be used on one scale—the one not in use—or the scales can be fixed together to make any length up to 13 in. radius, by the clamp shewn, and one head fixed on each as shewn in the figure. The scale clamp provides an adjustment of a most convenient kind, from 6½ in. to 13 in. radius; the spring centre leg can be adjusted $\frac{1}{16}$ in. or more by the screw which is seen, after the manner of a hair-divider. The beam compass pen leg is fitted with a handle so that it can be used as a ruling pen. The scales are eighths and twelfths.

CHAPTER V

LETTERING, FIGURING, DIMENSIONING, ETC., ETC.

Nothing in the way of ornamental work is ever done at the present time in ordinary Drawing Office practice.

Neatness, clearness, complete reference and description, making an unmistakable time-saving instruction to the workman, is all that is sought, needed or even tolerated. Ornamental work takes up time and costs money which nobody wants to pay, and has been absolutely cut out of modern drawing office work.

Rigid simple utility is what is aimed at and all which the commercial world is willing to pay for. It is only in exhibition drawing, which is used for striking the eye of the uninitiated, who may have something to do with giving orders, that anything approaching ornament is attempted. From any artistic point of view such attempts are usually failures and are best left alone. What is here shewn, therefore, is the minimum of ornament and the maximum of utility. Plain block Roman letters in forms which can be " printed " with a hand pen, and italics with all twists and tails eliminated, are all that are permitted, large titles or headings being made with stencil plates and the breaks joined by hand. The same remarks hold good for figuring. This should be large enough to be seen easily and clearly without being obtrusive and confusing to the general aspect of the drawing. Good hand printing of the kind named has a distinct value and should be assiduously practised by the beginner. A decent skill is quite necessary and is not to be considered at all optional. It is essentially a young draughtsman's art, and is seldom or never done by fully qualified draughtsmen whose time is expensive. In the young draughtsman's work it has a similar value to good line-making, it helps to put the best class of work into his hands, and, as has been said before, greatly enhances his chances of making progress.

In the *finish* of mechanical drawing, there is some analogy to art work, to the extent that in good finishing no part must obscure or interfere with the clearness of the rest.

The acquirement of a *free* style of hand printing often requires a good deal of time and practice, and the student is advised to carry out a systematic method which admits of frequent repetition of each single letter and figure. Strong, rough-faced, ruled foolscap paper is suitable for practice, with a surface as nearly resembling that of drawing paper as may be. Hand-printed lettering ranges from a little over $\frac{1}{8}$ inch high down to $\frac{1}{16}$, and both sizes should form exercises. Lines should always be ruled for the printing, especially on drawings, as distinguished from tracings, and, of course, all the same sizes for the same class of printing maintained throughout a drawing.

In commencing with letter exercises, it is a good plan to draw mechanically a whole line of one letter from the copy, p. 31; then leave two or more lines for exercise; then another line of the next letter, with more spaces for exercise; and so on, until the whole alphabet has been gone through. Certain letters are generally found to be difficult to certain hands, and extra practice in these is required. The same method applies to figures. It is not necessary to copy slavishly any particular style, for, as a matter of fact every man develops a style peculiar to his hand as in any other form of handwriting, and it is almost equally distinguishable. But the style must be clear and free and as mechanical as possible in general form, so as to correspond in character with that of the drawing. Peculiar "hieroglyphic" writing or printing is out of place on a mechanical drawing, where every line and letter should be unmistakable.

Examples of letters and figures are shewn on the opposite page.

ABCDEFCHIJKLMNOPQR
STUVWXYZ
1234567890

ABCDEFGHIJKLMNOPQRSTUVWXYZ
1234567890
1234567890

abcdefghijklmnopqrstuvwxyz

Side Elevation End Elevation Plan Cross Section
Vertical Section Wrot Iron Cast Iron
Brass Gun Metal

Fig. 10.

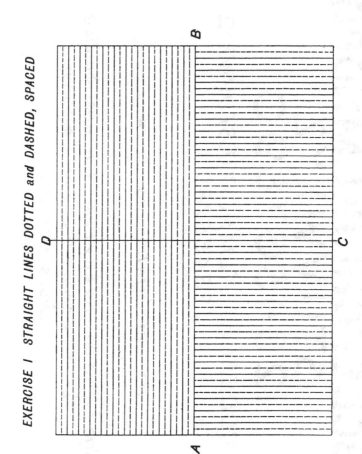

EXERCISE I   STRAIGHT LINES DOTTED and DASHED, SPACED

CHAPTER VI

EXERCISES

The purpose of the figure on the opposite page, EXERCISE I, is to give practice with the scale, pencil and ruling pen. It is to be drawn double the dimensions shewn in the book, the spaces shewn then being ⅛ of an inch wide. The drawing paper should be large enough to allow of the working of the T-square and set-square, without inconvenience from the pins in the corners of the paper, and the rectangular enclosing lines being 8 inches from side to side and 6 inches from bottom to top; quarter imperial size of paper will do; this will be 15 inches by 11 inches. The paper should be Whatman hot pressed or equal quality, and it should be pinned on the board without any puckers. A centre line *A, B* is then to be drawn across the paper with the T-square, the butt of the same being carefully held to the edge of the board. With a set-square, a vertical centre line (bottom to top) as *CD* is the next move.

At exactly 3 inches, set out by the scale and pencil on the vertical centre line, above and below the horizontal centre line, two horizontal lines reaching over 4 inches on each side of the vertical line are to be drawn with the T-square. The scale is then laid over the three lines at the 4-inch distances to make sure that they are parallel. If they are, 4-inch distances are to be laid out with the scale and pencil on each side of the vertical line, *on all three lines*, a division of the scale being laid dead true on

the vertical line, and the setting out being done by placing
the sharp edge of the pencil in the "cut" of the division
which occurs at the 4-inch distance.   As a test and guide
for holding the pencil square to the setting-out line, with
the sharp edges vertical, a trial on the "cut" of the centre
division and the vertical centre is to be made, before
marking the distances, the pencil being held firmly in the
fingers without alteration of position, during the marking.
The two vertical end lines are then to be drawn with the
"set-square," so as to "cut" the six set out distance points.
The 8 or 10-inch set-square would reach the whole distance
and the line could be drawn at one stroke, but if a short
set-square is used the line can be drawn by two strokes
"joined up," the lower part being drawn first and the
T-square and set-square then moved up to complete the
other part of the lines.   There is no objection to this as it
often has to be done in practical work, but the joining
must be done carefully, so as not to shew any break.   A
perfectly true rectangular figure of exact dimensions in
all directions should be the result.   The Exercise figure
shews horizonal lines in the upper part for practice with
the T-square, and vertical lines in the lower part for
practice with the set-square.   All the spaces are to be
exactly equal in width.   The lines are alternately solid
and "dotted": the dots or dashes are to be very close
together and the lengths approximately equal, as can be
fixed by sight (not measured), to shew uniformity and
continuity of line.   The spaces are to be "set out"
by means of the scale and pencil, the scale, of course,
necessarily having $\frac{1}{8}$-inch divisions, a "full-size" scale
being best if the student has one, although 3-inch or
$1\frac{1}{2}$-inch *fully divided* will do equally well.   A division on
the scale must be set "dead on" the centre line or on one

of the end lines of the rectangle, and *a* division will necessarily coincide with all three. The pencil is then tried into one of these to get its position in the hand which will mark off all the points required for the spaces correctly. The scale is held and the pencil manipulated in marking, as shewn in the photographic figure referring to scales (Fig. 6).

This Exercise is to be drawn in pencil, inked in with Indian ink and then traced on both paper and "cloth[1]." In ordinary practice, cloth tracings from which sun prints are made are the rule, and as it is much more difficult to trace on cloth than paper, it is advisable that the student should become accustomed to working upon it. It is not laid down as absolutely essential that cloth should be used, but that it is advisable. This Exercise is not an easy one to execute perfectly, but the student must repeat it until he can do it properly. The spaces when finished must be exactly equal and correspond with the marks on the scale, from end to end. The lines must be solid (not ragged) and they must be of uniform thickness throughout, so as to present a uniform tint when viewed from a little distance.

As the first step in the simplest operations of mechanical drawing, it is important that the student should submit himself to a rigid discipline of accuracy; this he must maintain throughout his studies in this subject. The ink lines must be relatively fine and placed dead on the centre of the pencil lines.

---

[1] Transparent tracing linen.

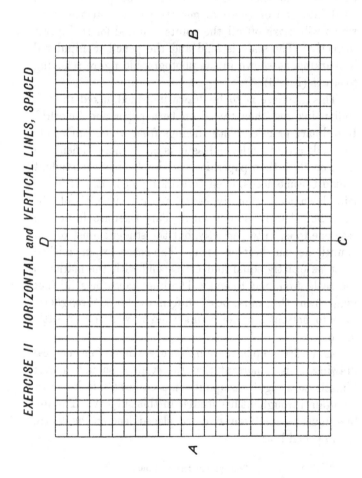

EXERCISE II   HORIZONTAL and VERTICAL LINES, SPACED

EXERCISE II is a combination of the two series of lines of Exercise I, and the general remarks preceding refer also to this. In all cases, *all* the spaces and intersections must coincide with the marks on the scale when laid across the drawing, both separately and collectively. Nothing short of visual exactness is to be passed.

In "inking in" the lines, in all cases in mechanical drawing, great care must be taken to get the ink lines central with the pencil lines, for, however correctly the pencil lines may be made, that accuracy may be spoiled by the ink lines not being properly placed. As the ink lines are always many times thicker than the pencil lines (which are extremely fine), a good deal of care and practice is required for the exact placing of the ink lines.

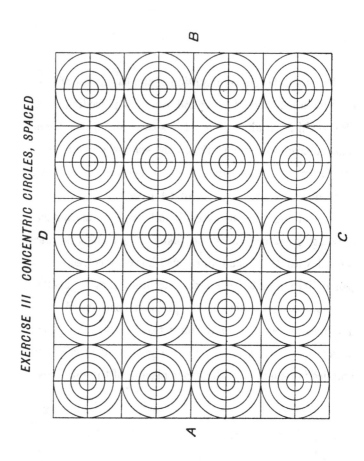

EXERCISE III   CONCENTRIC CIRCLES, SPACED

EXERCISE III contains a combination of straight line and compass work. It is to be drawn double the size shewn in the book and on paper at least quarter "imperial" size. As in the previous exercise a rectangular figure of exactly correct dimensions is first constructed. This is divided into squares of exactly equal size. Each alternate vertical line, the centre lines of the circles as well as the horizontal centre lines, are again divided into $\frac{3}{16}$-inch spaces for the sizes of the concentric inner circles. The intersections of the centre lines of the circles are then "pricked in" with extreme care, by means of the compass "centre" leg. The large circles are then drawn in so as to coincide exactly with the boundary lines of the larger squares in such a manner that the circle lines do not thicken the lines of the squares. The circles have thus to be made to fit the squares at four points—not at all an easy thing to do. The possibility of accomplishing this depends, first, on the accuracy of the squares, and, secondly, on the exactness with which the circle centres are pricked in on the intersections of the circle centre lines. There is no margin for any error anywhere. Of course all the circles will be made with the same "setting" of the compasses. When the large circles are correctly drawn, the smaller circles are made to cut the "set out" marks which have been made on the centre lines of the large circles.

It is not often that any of these early exercises are correctly executed by a beginner at the first trial, but it is imperative that they should be properly done before proceeding further.

This exercise is also to be drawn in pencil, inked in, and traced on paper and "cloth," or cloth only.

The final check on the accuracy of the work is made by laying a scale right across, when every line drawn must coincide exactly with the markings on the scale.

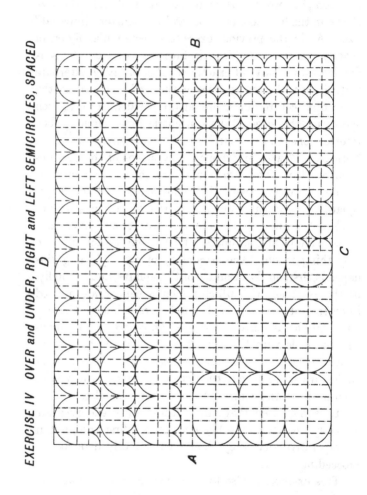

EXERCISE IV   OVER and UNDER, RIGHT and LEFT SEMICIRCLES, SPACED

EXERCISE IV contains further straight line and compass work combined. It is mainly intended for promoting the acquisition of skill in handling the " bows," both pencil and pen, with the drawing *fixed in position* on the board as in ordinary drawing office work; that is, without twisting the board around for accommodating the position of the hand. Drawing office boards are large, and for various reasons cannot be twisted around. The draughtsman therefore has to accommodate himself to the board, and has to draw curves "over" and "under" a line, and "right" and "left." These positions will be recognized in the figure.

The exercise is to be drawn double the size shewn in the book, when the rectangle containing will be 8 inches by 6 inches and all the spaces $\frac{1}{4}$ inches or multiples of that size.

All the distances and sizes are to be dead exact to scale, and all the curves are to fit the lines in three places and are to join each other exactly as shewn, without thickening the junctions.

In this case, also, the possibility of executing the figure depends upon the accuracy of the primary "setting out" of the straight lines and of the exact pricking of the curve centres in the intersections.

The centre lines here are dotted and dashed according to the ordinary custom of shewing centre lines on tracings for printing (sun printing), or on drawings where only black lines are used. On "inked in" drawings on paper, the centre lines are usually drawn in red colour, carmine or crimson lake, and dimension lines in blue colour.

The final check, again, is the scale laid right across the drawing. Uniformity in thickness of each kind of line, solid and dotted, is taken for granted.

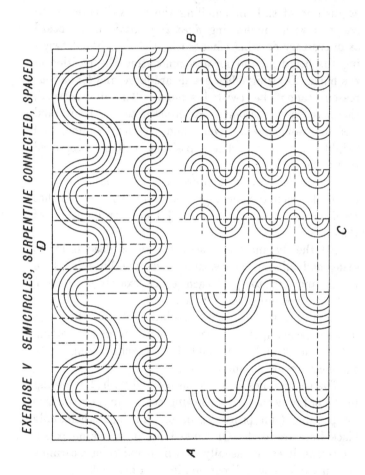

EXERCISE V SEMICIRCLES, SERPENTINE CONNECTED, SPACED

EXERCISE V is one of progressive difficulty over the preceding in straight and curved line combinations, especially the latter.

For successfully executing the junctions of the curves the primary setting out must be dead exact, as well as the pricking of the main centres.

The curves are to fit the straight lines as well as each other.

The exercise is to be drawn double the size in the book, making the whole figure 8 inches by 6 inches and the spaces between the curves $\frac{1}{8}$ inch.

The readiest method of setting out the exercise is to draw the boundary lines and then to set out $\frac{1}{8}$-inch spaces by the scale all round, drawing lines across to these after the manner of Exercise II, finally inking in the centre lines, as shewn, as well as the curves.

On account of the numerous points where the lines and curves have to fit exactly, great care is required in the primary setting out and at every point in the exercise.

As in the previous exercises, this is to be drawn in pencil, inked in and traced.

It may with advantage here be repeated that the ink lines of the curves must be as fine as practicable and consistent with expeditious work, and carefully placed central on the pencil lines. Great care must be taken to hold the compass square with the paper in drawing the curves, and to avoid "springing" the legs by any undue pressure. As a precaution against spoiling the whole figure by one wrong line, before every curve is made in ink, the junctions must be tried over by the compass ink point without making any mark, the line being made only when it has thus been ascertained that no accidental alteration of the legs has taken place.

No twisting of the drawing-board is to be permitted, and the final scale check, all over, is to be made.

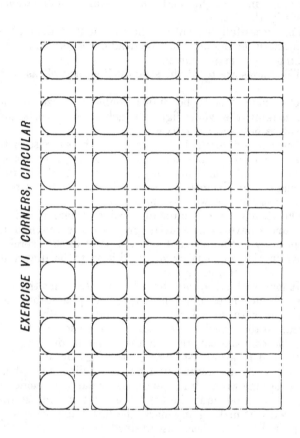

EXERCISE VI is designed for the cultivation of a ready skill in *putting in corners*, which so frequently occur in mechanical drawing, without geometrical setting out or *construction* lines. This operation is quickly and correctly accomplished, after having "set" the compass to the required radius from the scale, by placing the *pencil* leg on the line where the curve is to finish, then lightly placing the centre point in the position in the paper where (by sight) it seems likely to be. A trial with the pencil leg is then made to the line where the curve is to commence, and if the trial centre is in correct position, it is then to be pressed in so that it can be found again. The "corner" is then executed. If the trial centre is not correct the pencil point is returned to the line from which the trial started, and the centre point shifted and retried until the correct position is found. This method finds *one* distance from the "square" corner lines at once, and the second only has to be found by trial.

The exercise is to be drawn twice the scale of the figure in the book, and all the rectangular lines and positions are to be accurate as in previous exercises; the all over check of the scale to be also applied. The joinings of the corners and the lines to which they belong must be imperceptible.

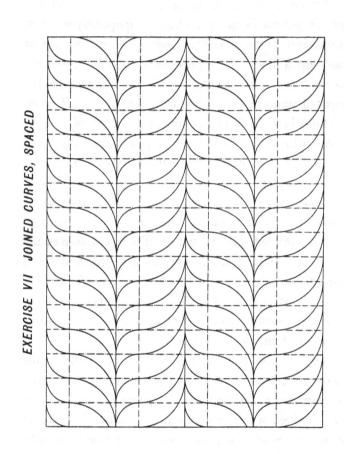

EXERCISE VII JOINED CURVES, SPACED

EXERCISE VII is designed for practice in the placing and joining of larger curves and with straight lines. These are relatively easier than some of the preceding and must be made rigidly exact both in the joinings and relationship with the straight lines, and the latter with one another.

EXERCISE VIII   OVALS

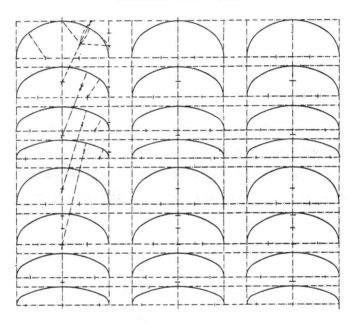

EXERCISE VIII is an useful example of curve combination, the conventional oval or ellipse. The particulars given are the length and half breadth, and something which will look about right is required, rather than a correct geometrical form. This refers, of course, to small scale drawing, in which an oblique view of any cylindrical part has to be shewn as an incidental. A large scale or size oval would have to be constructed, of course, by geometrical method.

A symmetrical oval can be made, of any dimensions, by drawing a major curve in a rectangle formed by the length and half diameter, so that it will touch the long side and cut the end lines at $\frac{1}{3}$ of their length as shewn in the first figure of the group. The radius for the end curve is found by trial. The result is what may be termed a decent oval, which, while easily made, is sufficiently correct in a mechanical drawing. The main points requiring care are the joinings of the curves which should be practically imperceptible. Finish is of more importance than geometrical accuracy.

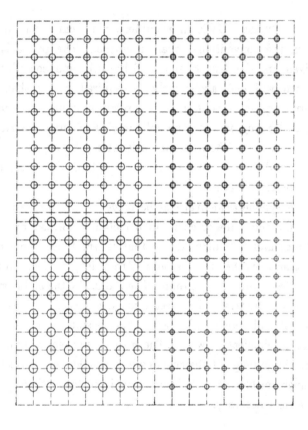

EXERCISE IX RIVETS and TUBES, SPACED

Rivets and Tubes 51

EXERCISE IX is in the pitching and drawing of small circles, such as rivets, stays, tubes, etc. The whole of the figure and centre lines are to be exactly set out as in the earlier exercises. The drawing is to be made double the size of the figure in the book, and inked in and traced. The circles would then be $\frac{3}{16}$, $\frac{1}{8}$ and $\frac{1}{16}$ inch diameter, and the tube set $\frac{1}{8}$ inch outside diameter, with the inside circle made as close to the outside as will leave a clear space between representing the thickness of the tube. The outside diameters in the whole series are to be correct to size, this being attained by taking off the radius from the scale as nearly as can be seen and then making a trial circle on a marginal space of the paper; this should be compared carefully with the scale as to its accuracy. The centre of the circle line is the *size*, especially when inked.

These circles can be made quite easily with the bow pen if that is properly set and adjusted. But spring bows may be used without departing from custom. Speaking generally, however, the larger bows should be used wherever they can be, as they are quicker in working, the adjustment being made usually by one movement when the draughtsman has acquired usual skill. Great care is necessary for "pricking in" the centres of the circles before commencing either to draw or ink the lines of the same. Expedition is important in such repetitions. All the circles are to be made with one setting of the compasses.

The dotted centre lines can be made by the pen of the bows with the centre leg bent up out of the way. The pen of the bows, speaking generally, can be adjusted to make a finer line than an ordinary ruling pen by reason of its smaller size. And for a similar reason spring bows' can be adjusted still finer. The term "fine line" is to be taken as a line made with a palpable opening of the pen nibs, and not as a black scratch which cannot be continued, and immediately runs dry as in the case of a close (quite close) nib. A line proper is one which will run indefinitely with uniform thickness from the pen.

The over-all scale test is to be applied at the finish.

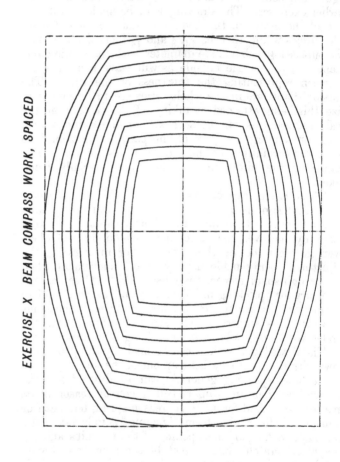

EXERCISE X   BEAM COMPASS WORK, SPACED

EXERCISE X is a series of curves to be made, at equal *centre line* distances, with one setting of a beam compass, or with the half set and lengthening bar, if the former is not available. In large work these latter instruments have to be used frequently. They are difficult to handle with precision, and it is necessary to have some practice with them.

The beam compasses are much the most satisfactory in working, though rather slower than the half set. In drawing curves with very long radii, the use of the beam compass becomes imperative unless "set curves" (curved rulers) are available. The centres for the curves shewn on the figure are all outside of the rectangle and can be made on slips of paper pinned on the board. The drawing is to be made double the size of the figure in the book, which makes the intervals of the curves cutting the vertical centre line $\frac{3}{16}$ of an inch wide, and those cutting the horizontal centre line $\frac{1}{4}$ inch wide. The radius for the former will be 6 inches and that of the latter 8 inches.

The scale check is to be here applied as before.

When the student has succeeded in drawing and finishing all the exercises up to this point and to the standard demanded, he will be able to make lines sufficiently well to commence object drawing. It is imperative, however, that he shall have done so, if he is ever to make a neat draughtsman, as by going forward before such standard is attained he is likely to develop bad habits of slovenly work, which will always tell against him. He is strongly advised, in spite of the tedium, to persevere until he *has* attained the standard; and there is no other way in which he can arrive at that point, more quickly and certainly, than the method here set before him.

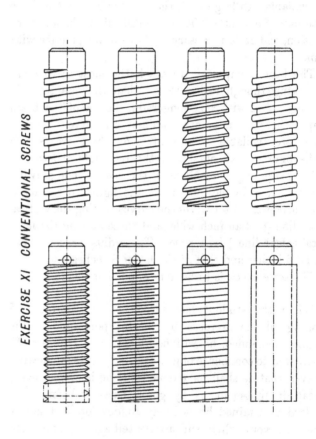

EXERCISE XI   CONVENTIONAL SCREWS

EXERCISE XI shews *conventional methods* of representing screws in mechanical drawing. None of these have any pretence to geometrical accuracy in the threads. In sizes in which threads are usually drawn, geometrical accuracy would be quite impracticable of execution, the process too slow and very much too expensive, and quite useless for practical mechanical purposes. Conventional forms have therefore had to be adopted, and these, being generally known, convey the meaning of the drawing quite clearly to the workman or whom it may concern.

Threads are said to be angular or square, in a general way, and for special purposes are sometimes made of different combinations of these. Worms intended to work with wheels come under different headings according to the form of tooth adopted. Pitches and particulars are written on the threads represented by conventional methods, according as they are Whitworth or American or other standard forms in the thread.

The outside diameters of the conventional forms, only, are shewn correctly. The bottom of the threads may also be shewn, but is seldom attempted, the forms and depths arising out of the dies, chasers or tools used for cutting.

The two left-hand top figures are such as are generally used for representing angular threads. The two lower figures of the same side may serve for either angular or square threads. The two upper figures on the right-hand are used for representing square threads; the third down is what is termed a "buttress" thread, used for presses; the lower is an usual representation of the thread of the leading screw of a lathe.

The series are for practice in drawing threads. They are "pitched" off a scale as before described and drawn with a set-square, as regards the cross lines, by inclining the T-square blade at a suitable angle, the blade being held by the left hand or set against pins in the board. The angular threads are generally drawn with a 60° set-square, although the Whitworth angle is 55°. Great care is required in setting out and drawing when actual threads are attempted.

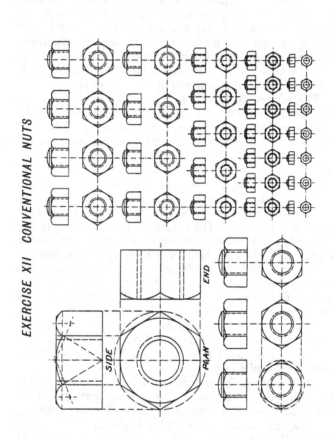

EXERCISE XII CONVENTIONAL NUTS

Nuts 57

EXERCISE XII is important, as the drawing of nuts is generally the only " ornamental " part of a mechanical drawing, and neatness is imperative. Without practice they are not easy to execute when the sizes are small.

The dimensions of nuts, as shewn on drawings, are *conventional*, speaking generally, and are necessarily so when small, as the exact standard sizes could not generally or practically be measurably shewn. *Conventionally*, nuts are shewn of a width across the corners of hexagon forms of twice the diameter of the bolts or studs on which they stand. The depths of nuts vary according to requirement, the conventional depth being one diameter of the bolt. Bolt heads are similarly represented, but are about ⅛th shallower. This kind of fitting, however, varies so much in everything but diameter of screw and pitch that particulars of depth are generally necessary. Nuts are seldom put into a drawing except for the purpose of making sure that they will go in and can be worked in the positions assigned. When they *are* shewn, however, they have to be well and properly drawn or they may spoil a whole sheet of work.

In the left-hand top corner of the figure a conventional nut is shewn full size and its projections to the different views are illustrated. In the " side elevation " the centre " flat " is shewn to be the diameter of the bolt and the side " flats " half that diameter. The curve for the top of the centre " flat " is drawn from a distance from the top line, which is equal to the width of the " flat," whatever the depth of the nut may be. An angle of 22½° (set out with the 67½° set-square) from the centre of the top of the nut cuts the centre line of the side " flats " in the proper point for the centre of the side top curves.

The usual way of commencing to draw a nut is to make a circle of the size of the " cross corners," and from the centre line of the nut to draw a 60° angle with the set-square. A circle is then drawn to touch the angular line. To this circle all the other " flats " are drawn, which completes the hexagon. The figure shews a hexagon single " canted " nut; it is obvious that a nut may be double " canted," that is, " canted " both top and bottom. Incidentally it may be said that " canting " is done to take off the corners of the nut flats and so form a circular " face," which can be turned in tightening up without hindrance from the corners digging into the opposite face. Also to avoid " burrs " from the corners getting accidentally damaged and for the sake of appearance and general convenience.

In drawing, even small sizes, it is best to get the radii of the curves by the method shewn in the full-size figure.

This exercise is to be drawn double the size shewn in the book.

EXERCISE XIII   SHADE LINING and SECTION LINING

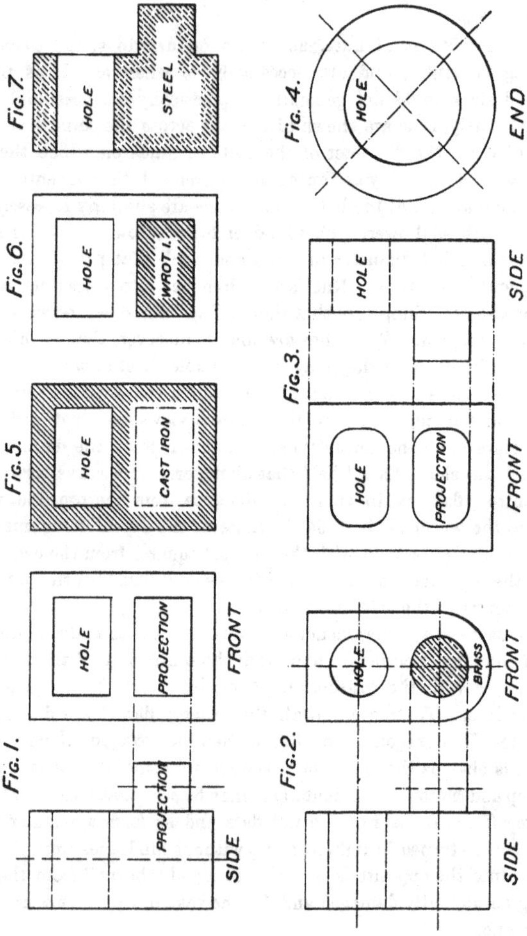

Fig. I.

HOLE

PROJECTION

FRONT

PROJECTION

SIDE

Fig. 5.

HOLE

CAST IRON

Fig. 6.

HOLE

WROT I.

Fig. 7.

HOLE

STEEL

Fig. 2.

HOLE

BRASS

FRONT

SIDE

Fig. 3.

HOLE

PROJECTION

FRONT

HOLE

SIDE

Fig. 4.

HOLE

END

EXERCISE XIII is for the purpose of illustrating conventional methods of indicating projections or holes, or the "dark" sides or edges of figures shewn on a flat surface. These consist of thick lines made on the side which would be in "shade," the light coming, in the supposition, from the left-hand top corner. Fig. 1 shews the side and front of an object in which is a square-cornered hole, and which has also a square-cornered projection. The dotted lines shew the direction of the hole from the side view, and the projection is also evident to the right-hand of the lower part of the same figure. In the front view a thick line is made on the upper and left-hand edges of the hole, these naturally being in "shade," while the bottom and right-hand side of the hole which would be "lighted" are shewn by fine lines. The "shade lining" of the projection is on the reverse sides to that of the hole. The whole figure in both views is treated in the same manner. Fig. 2 illustrates the method of shading round holes and round projections. As the darkest shade gradually merges into the lightest by reason of the curve, the "shade lines" are made to represent such an effect. In the shading of a cylindrical projection, it is not now customary to shew any "shade" on the under side, partly to make a distinction between a "round" and a "square" projection, and partly because the actual darkest part of the natural "shaded" side of a cylinder is not at the bottom of the figure, but some distance towards the centre line.

Fig. 3 is two views, front and side, of a similar form to Fig. 1, but both the hole and projection have rounded corners in the front view. The "dark" sides of the hole are thick lined to the round corners where these are tapered away to the fine lines of the "light" sides. The same method is used in the projection, but the lower side is dark lined to indicate that that side is "flat." Fig. 4 is the end of a pipe, the shading being a combination of that in the front view of Fig. 2, inside and outside.

The cross hatching in Figs. 5, 6, 7 and 2 is "section lining," as commonly used to indicate cast iron, wrot iron, steel and brass. Narrow areas, as in the upper part Fig. 5, are lined right across the section, while in the broad areas, as in the lower part of Figs. 5 and 7, the section lining is confined to the outer edges of a section.

Shade lining is seldom used on any but "show" drawings for non-technical or technical-commercial purposes. It takes time and costs money and is of little use to the technical mind, and it is a slow process by reason of the thick lines taking a long time to dry. Moreover it destroys the accuracy of the drawing to some extent, introducing doubt as to whether the inside or outside or centre of the line is the measure. At the same time it undoubtedly "clears up" the understanding or "reading" of a drawing, which is sometimes of importance.

"Section lining" is generally used on tracings for "printing," saving any further work on the print; but it is seldom done on inked drawings, colouring being preferred and more easily done; neutral tint being used for cast iron, blue for wrot iron, purple for steel and yellow for brass. The printed particulars on the sheet will assist these explanations.

The figures are to be drawn twice the size shewn in the book.

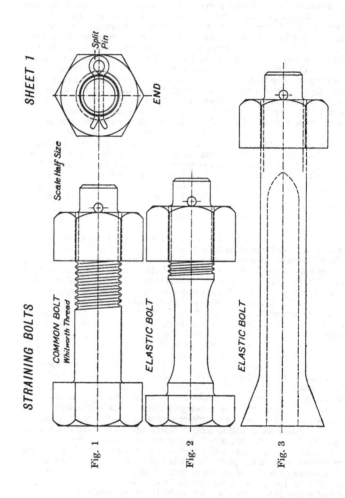

STRAINING BOLTS

SHEET 1

Scale Half Size

Split Pin

END

COMMON BOLT
Whitworth Thread

ELASTIC BOLT

ELASTIC BOLT

Fig. 1

Fig. 2

Fig. 3

CHAPTER VII

At this stage, a short series of practice drawings is introduced for further development of habitual methods of considering mechanical "views" and projections. They consist of general drawings of simple mechanical objects, the execution of which is intended as a preparation for the more complex ordinary "getting out" of sets of drawings for machines. The explanations of the subjects of these, and all other drawings following in this book, are mainly for the purpose of assisting the draughtsmanship aimed at, by a fair understanding of the nature and purpose of the object to be drawn, rather than for any technical instruction in the use of the objects. At the same time much may be learnt by the enquiring mind from such explanations.

SHEET 1 contains drawings of three types of fastening bolts. The figures are to be drawn double the size shewn in the book, which will make the diameters over the threads full size. The paper is to be 15" × 11". Fig. 1 shews the side elevation and end elevation of a "finished" common bolt and nut, with split pin for preventing the nut from working loose by vibration. The proportions are ordinary Whitworth, including the depths of the nut and head. The angle of the thread is drawn with a 60° set-square; the depth of the thread and the "rounding" of the top and bottom are about correct proportions. The object of drawing the thread is practice in small work. It would seldom or never be done in office work.

Fig. 2 is the side elevation of a tension bolt, which is designed to give elasticity and prevent fracture from shock. The reduced part of the shank has an area the ratio of which, to the area at the bottom of the thread, is equal to that of the limit of elasticity to the ultimate strength of the metal. Such a bolt could not be broken without stretching the reduced part about 20 per cent. of its length, and would consequently give notice of failure. The nut is deeper than ordinary, and such as is usually used on connecting rod end bolts for steam engines and similar work which is subject to heavy shocks.

Fig. 3 is an armour plate bolt; the object of the hole up the centre being the same as the reduction of the shank in Fig. 2.

The operations of drawing the objects always commence with drawing the centre lines in position; giving proper space for drawing, dimensioning and lettering.

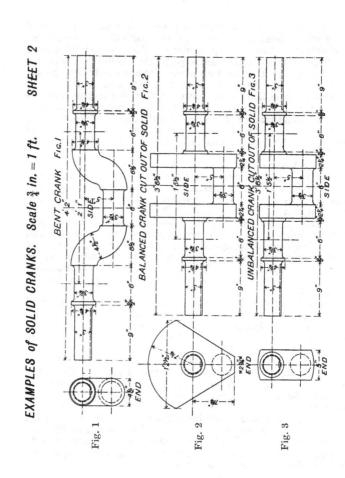

EXAMPLES of SOLID CRANKS.　Scale ¾ in. = 1 ft.　SHEET 2

Fig. 1

Fig. 2

Fig. 3

SHEET 2 contains drawings of three wrot iron or steel cranks for steam engines. They are of the double web type. Fig. 1 is of the "bent" type; Fig. 2 a double web balanced crank cut out of the solid; and Fig. 3 an unbalanced crank also cut out of the solid, that is, each of the latter are made out of forgings in which the crank portion is forged solid, the shaft ends being roughly forged near the size. The figures are to be drawn double the size of the figures in the book, or $1\frac{1}{2}''$ scale. Paper, $15'' \times 11''$. The smallness of the scale is chosen with a view to affording practice. There are many corners and joined curves which require careful drawing at the scale mentioned. All the distances are to be set out with the scale and pencil from the centre lines of the shafts and the vertical centre lines of the cranks. The shaft ends are shewn keywayed for fly-wheels or pulleys.

Cranks of this description would be "machined" all over, excepting the "bent" portions of Fig. 1.

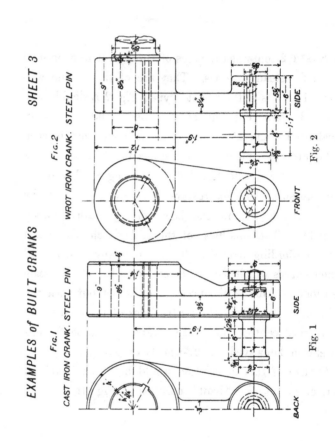

EXAMPLES of BUILT CRANKS   SHEET 3

FIG. 1   FIG. 2
CAST IRON CRANK. STEEL PIN   WROT IRON CRANK. STEEL PIN

Fig. 1

Fig. 2

SHEET 3 contains examples of built up single web cranks for steam engines. Fig. 1 shews a half end elevation and a side elevation of a cast iron crank web fitted with an iron or steel crank-pin. The crank shaft " eye " as shewn would be slotted with two key-ways, very slightly tapered from the outside; the main shaft body being " let in " to the crank boss about $\frac{1}{2}''$. The crank pin is made with a taper shank and a reduced end screwed and nutted for holding the pin in position. The crank web would be machined on the face and the backs of both the main boss and the pin boss, and chamfered all round the edges for a "finish." Fig. 2 shews a front elevation and side elevation of a wrot iron or steel crank, built up of a web with bosses and "fitted" pin of iron or steel. The crank would be forged roughly to size and machined all over. The crank pin would be prepared with the shank slightly too large for the hole in the boss, and the boss would be heated and "shrunk" on the pin. Or if put together cold, the pin would be forced into the crank by heavy hydraulic pressure. The two pin keys in the pin shank would be driven hard into parallel holes and then cut off close; they are for the purpose of preventing the pin turning round in the hole in case the connecting rod end should "run hot" and "seize." As in the former example the crank shaft collar is "let into" the main boss for the purpose of getting the crank journal close up and at the same time admitting of a strong collar for wear endwise. The drawing is commenced from the centre line of the crank shaft, and is to be made double the Book size on paper $15'' \times 11''$. The scale of the Book figure is $\frac{3}{4}''$ and that of the exercise drawing is to be $1\frac{1}{2}''$ to a foot.

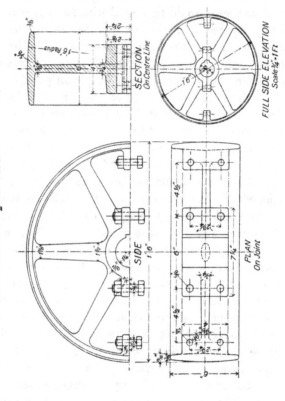

CAST IRON SPLIT PULLEY. Scale 1½ in. = 1 ft.
18 in. Diam. 6 in. Face 2½ in. Bore

SHEET 4

SECTION
On Centre Line

FULL SIDE ELEVATION
Scale ¾"=1 Ft

SIDE
1'6"

PLAN
On Joint

Driving Pulleys 67

SHEET 4 contains a side elevation, half cross section and "joint" plan, of a cast iron split pulley fitted with bolts, and a full side elevation of the same pulley to a smaller scale.

The Book scale of the large half view is $1\frac{1}{2}''$, and the exercise drawing is to be $3''$ to the foot, drawn on paper $15'' \times 11''$. The small scale figure would be $1\frac{1}{2}''$ scale in the exercise. The small circles, even of the exercise drawing, are rather small and will require careful drawing, but as the drawing is essentially an *exercise* for cultivating skill, no object would be gained by making it larger and easier to draw. The pulley would be described as an eighteen inch diameter, $6''$ face, $2\frac{1}{2}''$ bore, cast iron split pulley, put together with four $\frac{5}{8}''$ and four $\frac{1}{2}''$ bolts. At least two of the bolts would act as "steady pins" and be close fitted for that purpose.

The drawing is commenced on the horizontal and vertical centre lines of the pulley bore; and is to be finished strictly to the sizes shewn. The arm centre lines are drawn with the 60° set square. The taper of the arms may be made to a circle of the proper diameter made on the bore centre; the outer end of the taper being set out on each side of the arm centre on a circular line struck from the bore centre. This will bring all the corners in correct position. The same remarks apply to the central curves of the arms. The central and outer curves of the arms are made first and then joined with the arm lines.

5—2

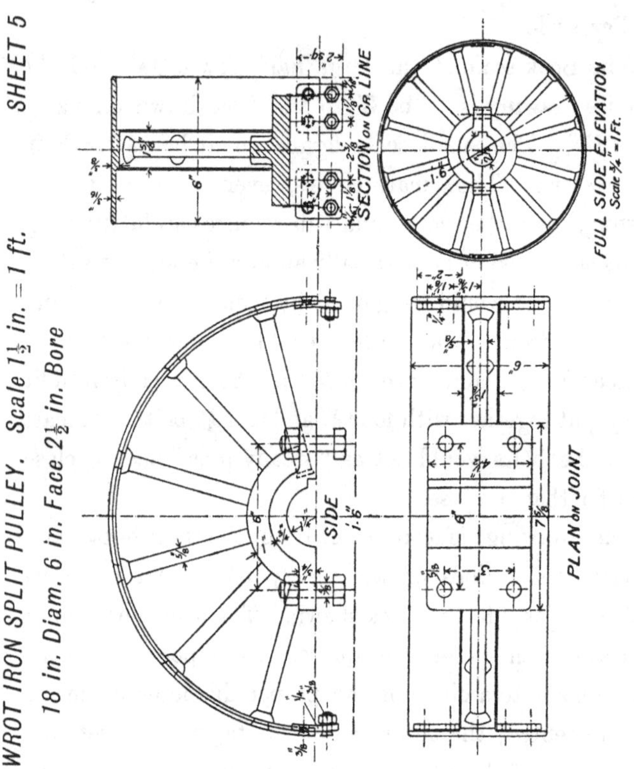

SHEET 5

WROT IRON SPLIT PULLEY. Scale 1½ in. = 1 ft.

18 in. Diam. 6 in. Face 2½ in. Bore

SECTION on CR. LINE

FULL SIDE ELEVATION
Scale ¾ = 1 ft.

SIDE
1:6

PLAN on JOINT

SHEET 5 is a drawing of a wrot iron split pulley. It would be described as an 18″ diameter, 6″ face, 2½″ bore, 12 arm, wrot iron split pulley, with cast iron boss, put together with four ⅝″ bolts and four 4¼″ joint plates.

The Book scale is 1½″ for the large figures of the half pulley and ¾″ for the whole pulley. The Exercise drawing is to be made on 15″ × 11″ paper, to 3″ and 1½″ scale respectively—double size of the Book figure.

All setting out of sizes is to be done with the scale and pencil as before described, and the "finish" is to be strictly to the sizes shewn. The drawing, at the scale, contains a good deal of small work which will not be easy. The twelve arm spaces are of course 30° each which may be set out by the 60° set square. The arms have to be divided into sixes on each side of the joint and consequently the half end spaces of the arms are 15°. By drawing the outside rim circle and laying off a 30° line from the centre, a space is obtained which can be divided by the compasses or even the scale, and so laid off from the centre line. A line can then be drawn at the 15°, right across the figure, and from two points on this line distances can be laid off with one setting of the compasses for a parallel line, on which the T square edge can be laid and from it the rest of the arm centre lines can be laid off with the 60° set square, from the pulley centre. The arms are of course parallel, made of round bar iron, and they are most easily drawn parallel by making a small circle of the right diameter on the centre line of each and then drawing the spoke lines in the same way as the centre lines, by means of the T square and set square. Care is required for preventing the shifting of the T square during the process, and two stout common pins stuck into the board close to the T square blade when in position (the butt being close to the board edge at one corner) is a useful way of retaining the T square in position while using the "set square."

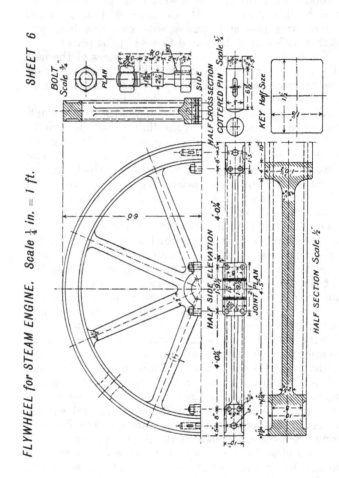

FLYWHEEL for STEAM ENGINE.   Scale ¼ in. = 1 ft.

SHEET 6

SHEET 6 is a drawing of half a cast iron flywheel for a steam engine or similar purpose.

The wheel is put together by eight $2\frac{1}{4}''$ bolts, and two $3''$, reduced to $2\frac{1}{2}''$, cottered pins, all of "Best Yorkshire" iron.

The wheel would be planed on the joint, all the bolts would be turned and placed in reamered holes, the rim lug bolts would be made "driving fits" to act as "steady pins," and the wheel would be bored, turned on the rim and edges and on both sides of the boss, with all fastenings tightened up. The cottered rim-pins would be made with the section of the cotter equal to half the section of the reduced centre of the pin, the two shearing sections of the cotter being thus equal to the section of the pin. The section of the pin round the cotter hole would be proportionate to the section of the reduced part, as the ultimate strength of the material is to its limit of elasticity. The cotter would be of hard Bessemer steel or "low" cast steel, thus securing an excess of strength over the "stretching" length of the rim-pin. Such a pin could not be broken short; it would, as "Best Yorkshire" iron, stretch 20 per cent. before breaking. The same remarks on proportion of bottom of thread section and reduced shank of all the bolts, would apply.

The wheel being 12 ft. diameter at $\frac{1}{4}''$ scale in the Book figure in the side elevation, vertical section and "joint" plan, and the enlarged section through rim, arm and boss being $\frac{1}{2}''$ scale, the exercise drawing which should be respectively $1''$ and $2''$ scale, would require half imperial size paper, or $22''$ long by $15''$ wide. The bolt and rim-pin drawings are $\frac{3}{4}''$ scale and the wheel key half size, on the Book figures; and these in the exercise would be $3''$ and full size respectively. There would be three low quality cast steel cotters in the wheel boss, two in one half and one in the other.

It would be more ordinary to draw the wheel $3''$ scale, but that would fill a double elephant sheet, and would make the detail rather large for practice.

SHEET 7

SPUR WHEEL. Made in Halves. Scale 1½ in. = 1 ft.

3ft.1in. Outside Diam. 50 Teeth. 2" Pitch. 8" Face. 3½" Bore.

CROSS SECTION

Joint Bolt

KEY

HALF SIDE

HALF PLAN on JOINT

bore

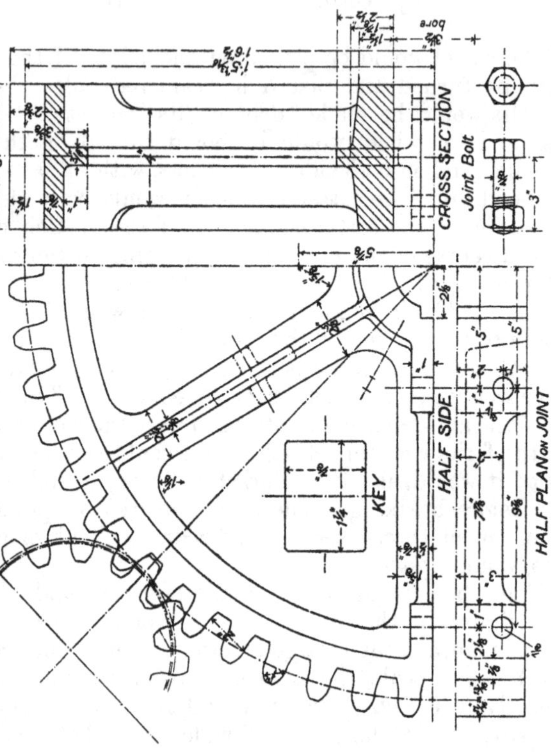

SHEET 7 is a quarter drawing of a spur wheel, and diagram of pinion teeth for the same.

The Book figure is a quarter side elevation, vertical half section and half of half the joint plan, the wheel being shewn as made in halves. It would be put together with eight $\frac{7}{8}''$ bolts at least two of which would be made to fill the holes in which they are placed, to act as steady pins. The wheel would be bored for the shaft, and keywayed, and as shewn would be rough machined for the joint faces. This practice drawing is to be made double the size of the Book figure, which would make the drawing 3″ scale, and as it is a good exercise in good class drawing, it may with advantage be drawn a full half wheel in both the side elevation and joint plan. The vertical section would be as shewn. This would require half imperial size paper 22″ × 15″. The centres of the radii of the teeth are shewn by dotted lines. The teeth are 2″ pitch and approximately cycloidal in form, and are drawn from the pitch line by two curves, one for the point (above the pitch line) and one for the root (below the pitch line).

All over the drawing careful work is required and very exact setting and handling of the compasses. The larger circles and tips and bottoms of the teeth are best drawn with a beam compass and short bar, if available. Otherwise the half set will have to be used, and special care must be exercised to avoid springing the legs. The half set is particularly liable to bore the centre hole large, which as far as possible is to be avoided by supporting the weight of the instrument by the left hand. This is awkward if the figure be "over" the centre, and turning the paper round for convenience in inking would be admissible. As there are 10 circles and arcs to be drawn from the centre and these have to be repeated at least twice in pencilling and inking, it will be well to execute the inner circles and arcs which can be done with the "bows" before risking the enlargement of the centre hole by the half set.

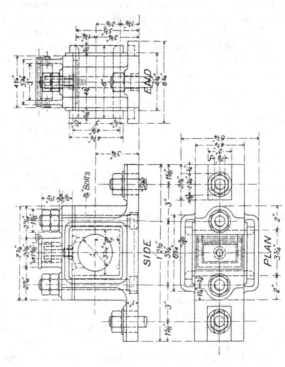

3" PLUMMER BLOCK. Scale 1½ in. = 1 ft.   SHEET 8

SHEET 8 is a general drawing of a 3″ plummer block to $1\frac{1}{2}$″ scale, and is to be drawn half size, or four times the size of the Book figure. 3″ plummer block, means a bearing for a 3″ shaft. The "block" casting or base is of cast iron as also is the "cap," and the oil box is part of the casting of the latter; the "lid" of the oil box is of brass, mounted by means of a brass frame or rim round the inside of the sides of the oil box, which carries the hinges. The lid is formed with a "heel" which prevents the lid from falling right over on to the holding down bolt ends. The lid frame is secured in position by two small screws with "pin" ends. The "wick tube" of the oil box is part of the iron casting of the "cap," and a short brass tube carries the oil clear of the cap so that it drops on to the hollow of the upper surface of the actual (brass) bearing. A recess is made round the oil hole so that a projection is obtained for the brass oil tube or ferrule for leading the oil "clear," without any projection from the planed cap joint. The actual "bearing" for the "shaft" is termed the "brass," and is made of brass, gun metal, phosphor bronze, or "white metal" as the case requires. It is carefully "fitted" into a machined "gap" in the block casting, and projects on both sides of the block to extend the "bearing surface" of the shaft. The hollow projections from the block casting at each side, under the "brasses," are "trays" for catching the oil "drippings" from the journal. The large bolts securing the cap are termed the "cap bolts," and the bolts securing the "block" to any "bracket" or "seat" are the "holding down" bolts.

All the sizes in this Exercise are to be "scaled" off the Book figure. $\frac{1}{8}$ of an inch actual will (in the Book) represent 1″, $\frac{1}{16}$″ half an inch, $\frac{1}{32}$″ a quarter of an inch, and $\frac{1}{64}$″ ($\frac{1}{32}$″ divided by sight) will represent $\frac{1}{8}$ of an inch actual. There are no $\frac{1}{16}$ inches in the figure, such small dimensions having been avoided in order to leave no doubt in the mind of students as to what the sizes are intended to be. The drawing will be made on half imperial size paper. All principal measurements are to be made from the centre lines of the figures.

CHAPTER VIII

8″ LATHE DRAWINGS

At this stage it is assumed that the student is able to "draw" *mechanically*, or by mechanical methods, and that it is now time to apply his skill to useful purposes, by the usual methods adopted in ordinary drawing-office work, as no other applications have any appreciable commercial value.

For this purpose it will be necessary to "get out" a machine in such ordinary way; and with a view to the general usefulness of the knowledge gained in the process of "getting out" a set of drawings, a complete lathe, with "sliding," "surfacing," and "screw cutting" motions and appliances, has been adopted as the subject. While mainly consisting of *wheel, screw,* and *sliding* motions, which are chiefly *matters of motion,* and rigidity, and the proportions arising more out of experience than calculations, it is a good general example of a piece of mechanism, the draughting of which affords opportunities for the exercise of skill and improvement of the same; and lends itself well in a general way to an understanding of "group design" and combination as well as to the sufficient illustration of drawing-office methods.

Such a lathe as forms the subject of the work following this point in the course of instruction, consists of several distinct groups of parts, each of which performs some specific part of the work of the lathe or is essential to such performance.

In the set of drawings following, the parts are grouped under the headings of: the Fast Head, which contains the motions for rotating the piece of work operated upon and provides *motion* for all the other groups; the Loose Head, which supports the outer end of the " work " by means of a stationary "centre" for which it provides adjustments; the Saddle and Slide Rest, on which the cutting tools are fixed; the "End Motion," which provides the "speeds" and driving actions for sliding, surfacing and screw-cutting; the Leading Screw and Back Shaft, which give motion to the "Saddle" parts; the Gantry, which carries and connects the whole of the other parts; and finally, a General Drawing of the assemblage of the whole of the groups as a complete Lathe.

The Example adopted is an 8″ centre strong lathe suitable for dealing with any kind or weight of work which can be "got in" over the saddle—a maximum of 11 inches in diameter; or of surfacing or otherwise dealing in the " Gap," with any article which will go into that, up to 32 inches in diameter. The lathe is especially designed for cutting heavy "worms" up to 2″ pitch of thread, and is provided with change wheels, which will "cut" any pitch of "eighth" measurement, from 2 inches down to $\frac{1}{16}$ of an inch, and the whole range of the Whitworth Standard threads including the gas threads. The " change wheel " teeth are extra strong, to provide for heavy " worm " cutting, and the numbers are made to avoid very large diameters of wheel. The " Banjo " or change wheel frame is made to take all the combinations which are necessary for the above pitches.

The general plan of the arrangement of the drawings is made on the assumption of no knowledge whatever of the parts of the lathe, on the part of the student; and all the details and combinations are laid out to be as clear as possible, as true mechanical representations of pieces of mechanism in the first place; and in the combinations, to provide what general instruction can be derived from the conventional execution of a Mechanical Drawing of the particular subject, without any special reference to any mathematical features; or of any practical uses to which such a machine may be put.

The Book deals with a system of teaching Mechanical Drawing, to the end of Drawing-Office conventional practice, which in itself is sufficient to form a separate branch of study of commercial value, for the young draughtsman.

In carrying out the general plan of the work in this Chapter, " model arrangements " of the general drawings of the separate groups of parts are made in outline (with no details). These models are for the purpose of setting out the centre lines of these general drawings so as to leave proper space for the figures, and for symmetrical arrangement of these on the size of sheet on which it is intended they should be drawn. The detailed views for these general drawings are shewn to larger scales, on subsequent sheets, and the complete dimensions and details of parts of the detailed views are shewn on further sheets which correspond to ordinary shop drawings. In this way "general drawings" of the separate groups are made, with every detail and particular provided. The "shop drawing" sheets are made in separate "collections" of parts, as they are ordinarily issued to the various shop departments: so for patterns, castings, forgings, &c., and also for machining of the parts of the groups.

In this way, a novice in the particular class of machinery to which the drawings refer is enabled to produce a full set of drawings in orthodox fashion, and by the process acquire a knowledge of ordinary methods of "getting out" work, as well as, incidentally, learn a good deal about the possibilities of the machine itself.

Explanations of the parts and purposes of the figures in the various sheets are given for the purpose of enabling the student, by an intelligent appreciation of them, to draw in such parts, in the general drawing, more correctly in the minutiae, than he would be likely to do, knowing nothing of their purpose or action.

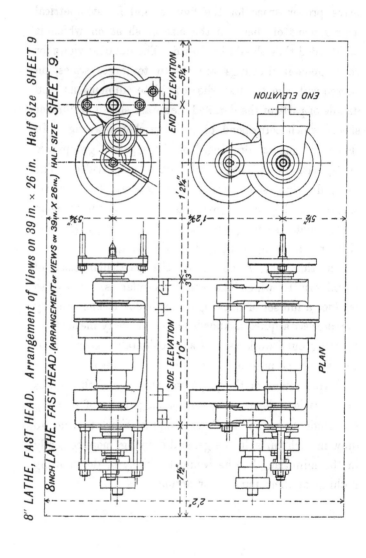

8" LATHE, FAST HEAD. Arrangement of Views on 39 in. × 26 in. Half Size SHEET 9

SHEET 9 is an outline drawing of the four views of the Fast Head of the lathe, such as are usually made to shew the whole of the parts and fittings. The dotted border line is the size of a double elephant sheet of drawing paper, trimmed round the edges, leaving $39'' \times 26''$ available space. The Book figure shews the *paper*, $1\frac{1}{2}''$ scale; while the *lathe head* being half size is $\frac{3}{4}''$ scale.

The centre lines of the drawing are to be laid out on the paper, but the figures or views are to be filled in from Sheets 10, 11 and 12; and for any small dimensions which cannot readily be made out with the scale reference is to be made to Sheets 14, 15, 16, 17, 18 and 19. These supply full particulars.

It will be seen that Sheets 10, 11 and 12 are not to be drawn separately as they are already in the general drawing, Sheet 9. The general dimensions as shewn in Sheets 10, 11 and 12 are to be put into the general drawing.

The principal parts of the Fast Head are the "head casting" as shewn in four views in Sheet 13, and in full detail dimensioned, in Sheets 14 and 15; the "spindle," shewn in Sheet 18; with spindle "bushes" or bearings in Sheet 19; and "adjusting nuts," "tail studs," "cross bar," "tail wheel" and "centre," shewn in Sheet 18. These should be drawn 'in first in the side elevation, the body of the spindle and parts which come inside the head casting being dotted, as shewn in Sheet 10. The spindle spur wheel, which is keyed on the spindle, is next drawn, and after that the spindle cone pulleys and pinion as shewn in Sheet 16, which are " loose " on the spindle, but which can be made fast to the spindle " spur wheel " as occasion requires, by the engaging or coupling clamp shewn on Sheet 19. With the " tail adjustment " in Sheet 18 and the " catch plate " and studs, the " spindle line " is then complete.

The Plan of the Fast Head is the next step, and all the parts so far shewn in the Side Elevation are to be drawn into the Plan. The Back Gear is then to be filled into the Plan, from Sheet 19 in which the back gear spindle and eccentrics are shewn, and Sheet 17 in which the back gear barrel and pinion and spur wheel are shewn. The " changing lever " or " spindle lever " from Sheet 19 completes the " back gear " in the Plan. A Stud, with spur pinion, engaging with the spindle " tail wheel," and three stepped cone pulley for driving the " back shaft " of the lathe, completes the fittings of the Fast Head, with the exception of a stud in the back end for clamping the "Banjo," and the clamp screws and centering screws which are concerned in fixing the Fast Head into the Gantry or Bed of the Lathe.

The Back Gear may now be projected to and drawn in the Side Elevation, which with the addition of the " catch plate " and studs completes that view.

The End Elevations, front and back, are projected from the Side Elevation and Plan respectively, particulars of the centres being also obtained from Sheet 15.

The Back Gear is for the purpose of reducing the speed of the lathe spindle and rate of turning of the work, at the same time increasing the force of the turning power. The eccentrics at the ends of the back gear spindle, by means of the hand lever, throw the gear in or out of engagement as required, the positions being secured by the two headed-pins which pass through the bosses of the head casting back gear brackets, and engage in suitable holes provided in the eccentrics. When the back gear is to be used the engaging clamp of the spindle spur wheel is disengaged from the spindle cone pulley. The Pulley, then, by means of its pinion, drives the back gear spur wheel, while the back gear pinion drives the spindle spur wheel and the spindle, to which it is keyed, thus rotating the work about nine times slower than the cone.

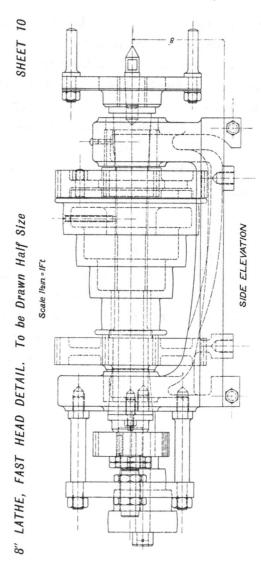

8" LATHE, FAST HEAD DETAIL. To be Drawn Half Size

Scale 1½in.=1Ft

SHEET 10

SIDE ELEVATION

To Be Scaled Off or Drawn From Detailed Parts

8 *in. Lathe, Fast Head* 85

SHEET 10 is the Side Elevation of the Fast Head, and
contains all the parts of the same in position. The parts
are named in accordance with the expressions used for
Sheet 9 as the "general drawing."

The head end of the spindle is coned and case hardened,
and ground very true, and it is mounted in a gun-metal
or phosphor bronze bush or bearing which is provided
with a key for preventing its turning in the hole in the
casting in which it is mounted, in case the bearing "runs
hot." The bush may be turned bottom to top, when the
bearing wears "down." The taper is for the purpose of
taking up any wear of the spindle or bush hole; it is "set
back" by "taking up" the nuts at the back of the tail
wheel and adjusting the "adjusting piece" which is
mounted in the cross bar, so that the cone of the spindle
just fills the hole in which it runs without "jamming"—
running tight.

The "tail end" of the spindle is a parallel fitting, also
case hardened and ground; and runs in a gun-metal bush
which is made in three parts, so that by "easing" the
joints they may be made to "close in" on the spindle and
take up any wear. The three part bush is mounted in a
taper hole in the head casting, and is held in position and
prevented from turning in the hole by a flanged collar
having "joggles" or solid projections which engage in the
parts of the bush. The tail wheel runs against this collar
and prevents the spindle from getting "forward" and so
slacking the fit of the head cone. The tail wheel is in
turn held in position by the lock nuts. The tail adjustment or pressure piece takes the end thrust due to work
and the pressure of the loose-head centre, and prevents
the "jamming" of the head cone. The catch plate, by
means of the studs, turns or rotates the "work."

SHEET 11

8" LATHE, FAST HEAD DETAIL.　To be Drawn Half Size

Scale 1½in=1Ft.

PLAN

To Be Scaled Off or Drawn From Detailed Parts

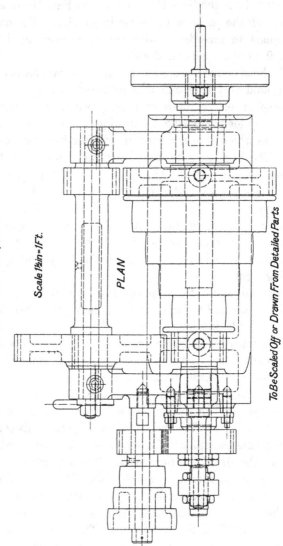

SHEET 11 contains the Plan of the Fast Head, in which the Back Gear is clearly indicated, and is "out of gear" in the position in which it is drawn. The Back Gear spindle or shaft goes right through the "barrel" of the gear and the eccentrics, and is keyed into the "head end" eccentric, as well as into the tail end eccentric; and is held in its position in the head casting brackets by the boss of the Back Gear spur wheel and the "plate" of the "hand lever," these overlapping the hole in which the tail eccentric is mounted, and bearing against the faces of the back end bracket boss. The Back Gear spur wheel is keyed on to the "barrel," the B. G. pinion being "cast on" the same. All the wheel teeth in a good lathe would be machine "cut." The B. G. when in action rotates on the B. G. spindle which is stationary. The semi-rotation of the eccentrics and spindle engages or disengages the wheel and pinion teeth of the Back Gear and those of the spindle and cone.

The driving of the stepped pulley or cone of the Back Shaft or Traversing Motion, mounted on a stud in the Head Casting, is clearly shewn: the "tail wheel" engaging in a wheel (changeable) which is keyed on the pulley boss.

The arrangement of the collar adjustment, with studs for the "spindle" tail bearing (split bush), is to be noted. The bosses for the fixing screws (clamps) of the "fast head" casting are plainly seen on the centre line.

The remainder of the parts have been explained in connection with Sheet 10.

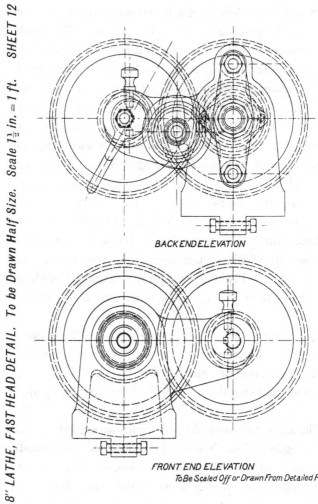

8" LATHE, FAST HEAD DETAIL.   To be Drawn Half Size.   Scale $1\frac{1}{2}$ in. = 1 ft.   SHEET 12

BACK END ELEVATION

FRONT END ELEVATION
To Be Scaled Off or Drawn From Detailed Pts

SHEET 12 contains the Back End Elevation and the Front End Elevation of the Fast Head. These are for transference to the General Drawing, and are mainly useful in clearly shewing the positions of the " centres " of the Lathe Spindle, the Back Gear Spindle and the eccentrics belonging to the latter. Incidentally, also for those of the traversing cone stud, the tail bearing collar studs, and the " tail " studs carrying the tail cross bar. The " side " adjustments for the "head" casting, in the "top table" centre gap of the Gantry or Bed, are also clearly indicated.

The hand lever for operating the Back Gear eccentrics is to be noted. It is made with a circular plate which covers the end of the tail end eccentric, and is secured by a nut on a continuation of the B. G. spindle ; being prevented from rotating on the spindle end by a projection of the eccentric key into a suitable key-way made in the hand lever plate, the nut covering the key end. The eccentrics are secured in their " in gear " and " out of gear" positions by circular pins (locking pins) which pass through the bracket bosses into holes corresponding, made in the eccentrics. Both front and back elevations shew this very clearly.

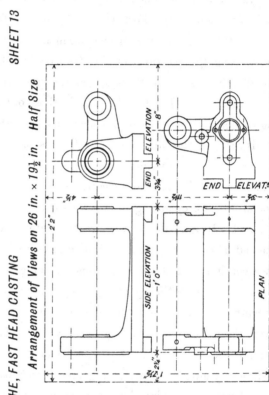

SHEET 13

Half Size

Arrangement of Views on 26 in. × 19½ in.

8″ LATHE, FAST HEAD CASTING

END ELEVATION

END ELEVAT.^N

SIDE ELEVATION

PLAN

8″

3¾″

4½″

2′2″

11½″

3½″

1′0″

2¼″

1′2½″

SHEET 13 contains an outline arrangement for the positions of views in a detail drawing of the Fast Head Casting, to be drawn "half size" on a half double elephant sheet of paper, or on half the space on such a sheet, in case it is elected to "get out" also the Loose Head Casting drawing on the same sheet, and so arranged that the tracings or prints can be cut up if desired for workshop use.

The figures are not to be drawn as shewn, but are to be fully detailed from Sheets 14 and 15. The dotted border line represents the space of a half double elephant sheet of paper to $1\frac{1}{2}''$ scale in the Book, the scale of the figures being $\frac{3}{4}''$ to a foot. The paper will of course be full size, $26'' \times 19\frac{1}{2}''$, or eight times the actual dimensions in the Book, the figures being also eight times the Book dimensions will be half size in the drawing.

Such a drawing as this sheet shews is made for the convenience of the workman in preparing the patterns for the metal castings in the first place, and later for the use of the machine man in preparing the casting for the reception of the parts which are to be mounted in or upon it. Separated from the rest of the parts in this way the features and machined surfaces can be more clearly indicated to the workman than by the more complex General Drawing of the Head. Every feature and machined surface should be shewn, described and dimensioned in this drawing.

This Sheet can be drawn on a double elephant sheet as before stated, and the remainder of the space of the paper can be filled in from Sheets 16 and 17, thus making a complete "pattern" drawing for the "pattern" shop. The print of this can be cut up into the respective groups as shewn, if desired, either for making the wood patterns or for machining the castings.

8" LATHE, FAST HEAD CASTING DETAIL   SHEET 14

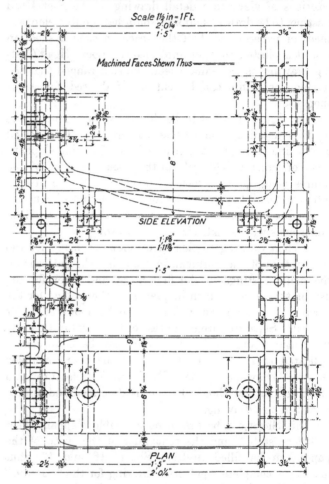

SHEET 14 contains a Side Elevation and Plan of the Fast Head Casting, fully dimensioned to finished sizes. These views are given separately, to give space for the dimension letters and figures, in the available size of page in the Book, and to avoid folded pages which are liable to get damaged by use.

Various devices have been used for indicating machined surfaces on drawings; red lines, dotted lines and so forth, but as in the present day practically all drawings are "blue" printed, and as it is undesirable to have anything to do on the prints after they are made, it is customary to "write" and use "pointing lines," on the print tracings, to a sufficient extent to explain what is to be done, without going to the trouble of red-lining the machined surfaces on the print. Moreover, in most classes of work, the various workmen are sufficiently well acquainted with the purposes of the drawings, to know which are machined surfaces and which are not. The general term "Finished Sizes" is ordinarily sufficient to indicate that rough castings or forgings must have sufficient metal "left on" for leaving good surfaces to the dimensions shewn, after the machining is done.

On the whole, as a general practice, it is better to indicate plainly which are machined surfaces, than to leave any doubt on the point; the draughtsman being held responsible, nominally, for the accuracy of the finished article.

The "finish" of this sheet, generally, indicates the nature of the principles which are adopted in getting out working drawings, and ensuring accuracy of execution of work, without oral directions to the workman. If both draughtsman and workman properly understand their respective businesses, the drawing should be sufficient to convey all the instructions necessary. Individual firms have their own particular methods of doing things, which are generally what they consider the cheapest, and sufficient for their purpose. At the same time it is necessary for the draughtsman to be brought up to a complete method, so that he is "competent" in any circumstances.

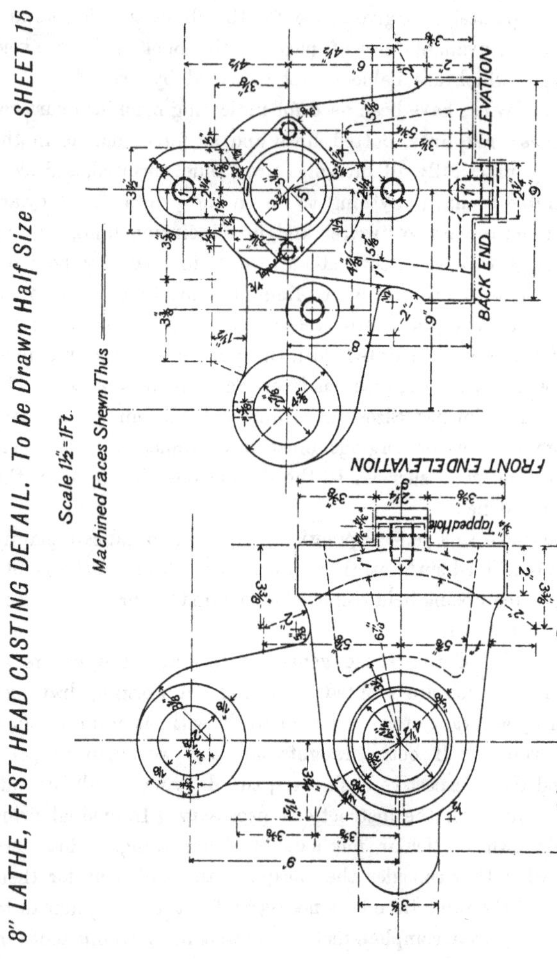

8" LATHE, FAST HEAD CASTING DETAIL.　To be Drawn Half Size　SHEET 15

Scale 1½"=1Ft.

Machined Faces Shewn Thus ⎯⎯⎯⎯⎯

FRONT END ELEVATION

BACK END ELEVATION

SHEET 15 contains the Front and Back Elevations of the "Fast Head Casting," for transference to the drawing of the Fast Head Casting according to the arrangement shewn in Sheet 13.

All the boring, facing, planing, drilling and tapping is shewn; also the centres and radii of the various curves.

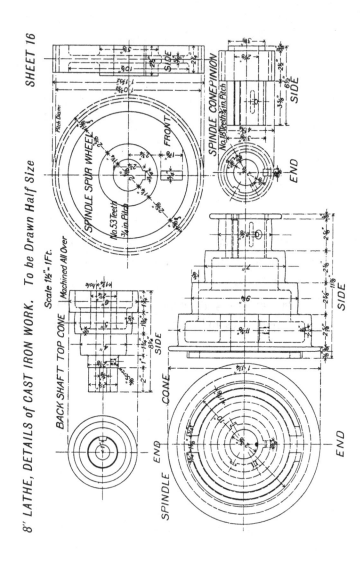

SHEET 16

8" LATHE, DETAILS of CAST IRON WORK.　To be Drawn Half Size

Scale 1½"=1Ft.

SHEET 16 contains detail drawings of the Spindle spur wheel, the Spindle cone pulley and pinion, and the Traversing Motion cone pulley. These are all cast-iron elements, and all the teeth are cut.

The clamp slot for the coupling bolt, in the spur wheel, and the lugs on the face of the cone pulley for engaging the coupling bolt are to be noted, as well as the method of putting together the cone and pinion. This is done by three shallow keys, to make a strong connection, also on account of the thinness of the small "step" of the pulley.

The pinion belonging to the Traversing Motion cone pulley being wrot metal and changeable is not shewn on this Sheet, but the boss for mounting it will be noted.

These parts are also to be transferred to the General Drawing of the Fast Head.

They are also to be drawn on the pattern sheet, as mentioned in connection with Sheet 13, in a group which can be used separately by cutting up the print.

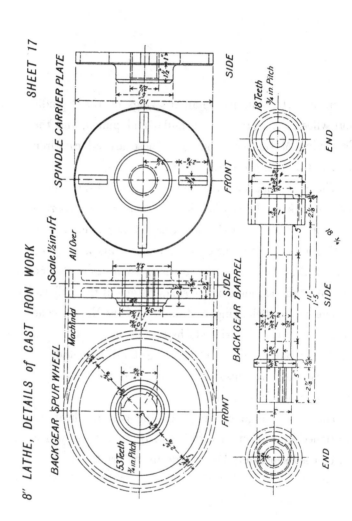

SHEET 17

SPINDLE CARRIER PLATE

8" LATHE, DETAILS of CAST IRON WORK

Scale 1½in=1 Ft

All Over

BACK GEAR SPUR WHEEL

Machined

SIDE

FRONT

BACK GEAR BARREL

53 Teeth
¾ in Pitch

18 Teeth
¾ in Pitch

END

SIDE

END

FRONT

SHEET 17 contains drawings of the Back Gear "spur wheel" in two views; also the "catch or carrier plate" casting, and the B.G. "barrel." These are to be referred to when making the Fast Head general drawing. They are also to form a group in the F.H. pattern and machine work sheet as before mentioned.

Both spur wheel and pinion teeth are to be "cut." The pinion, as is seen, is cast on the "barrel." All these parts are machined all over excepting the back of the "catch plate."

The gun-metal "tail bush" and collar for the Fast Head on Sheet 19 may be included in the pattern sheet or drawing, so placed that the print (which would be of the whole drawing sheet) could be "cut up," and these "brass fittings" given out to the brass fitter, separately.

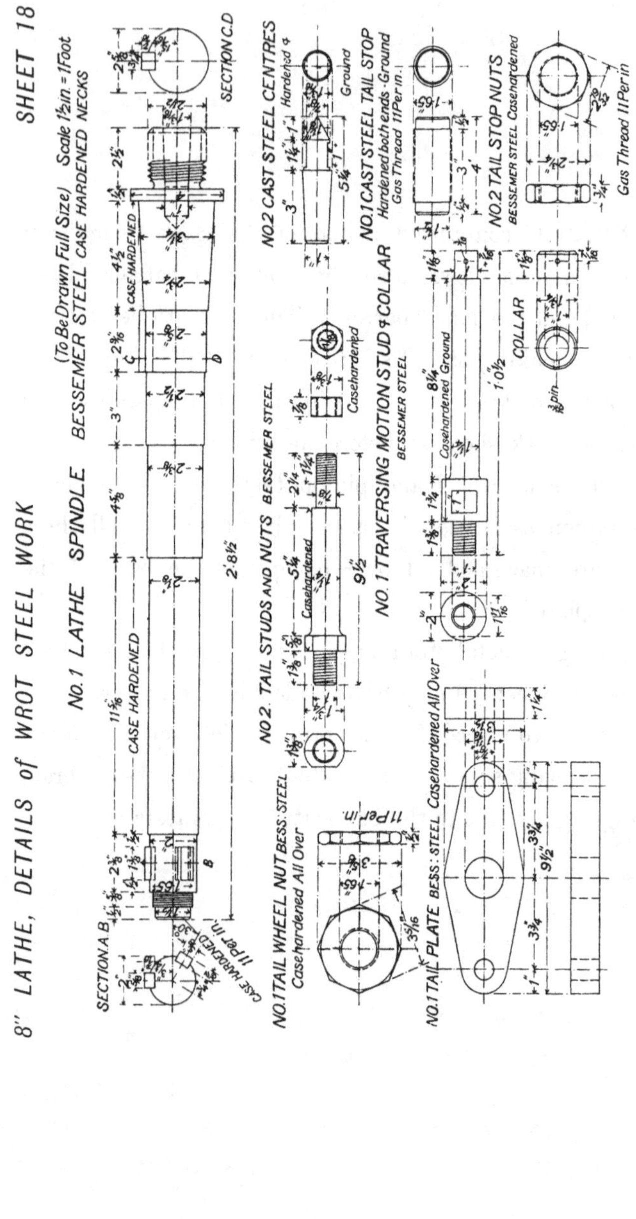

SHEET 18

8" LATHE, DETAILS of WROT STEEL WORK

NO.1 LATHE SPINDLE   BESSEMER STEEL CASE HARDENED NECKS

(To Be Drawn Full Size) Scale 1½in = 1Foot

SECTION C.D.

CASE HARDENED

SECTION A.B.

NO.2 CAST STEEL CENTRES
Hardened
Ground

NO.1 CAST STEEL TAIL STOP
Hardened both ends Ground
Gas Thread 11 Per in.

NO.2 TAIL STOP NUTS
BESSEMER STEEL Casehardened

Gas Thread 11 Per in.

NO.2 TAIL STUDS AND NUTS BESSEMER STEEL
Casehardened

NO.1 TRAVERSING MOTION STUD & COLLAR
BESSEMER STEEL
Casehardened Ground

COLLAR

NO.1 TAIL WHEEL NUT BESS: STEEL
Casehardened All Over

NO.1 TAIL PLATE BESS: STEEL Casehardened All Over

SHEET 18 contains the Lathe "Spindle" and all its steel fittings.

All these parts are to be examples of the best class of mechanical work. The spindle should be dead straight and true in all parts and be case hardened and ground on the bearings, as also on the fitting for the "tail" wheels. All the keys should be unhardened cast steel. There are five "tail wheels" to give proper counts for the wide range of screw cutting and traversing which is possible.

The spindle "centre" piece should be of cast steel, hardened at the point and ground true. It is imperative that the "centre" hole in the spindle head should be dead true (central) throughout its length, so that the "centre" can be ground true at the end in position, and can be turned round in the hole without being out of truth in any position.

This Sheet and part of Sheet 19 combined on one drawing would be such as would be got out for the smiths' shop where the necessary forgings are made, and would be suitable also for the machine shop where the articles are "finished."

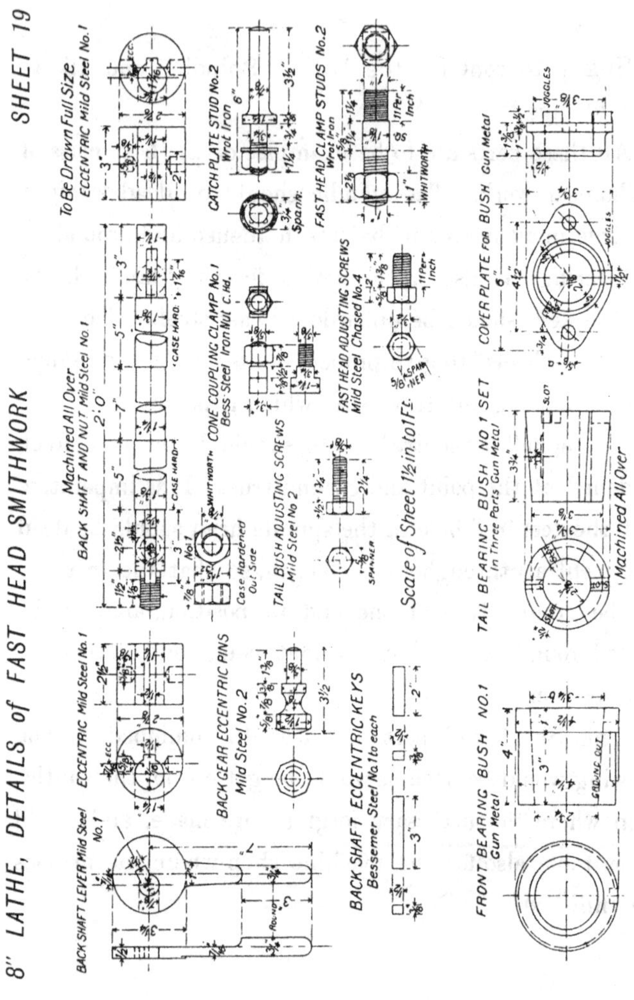
8" LATHE, DETAILS of FAST HEAD SMITHWORK
SHEET 19
To Be Drawn Full Size
Machined All Over
BACK SHAFT LEVER Mild Steel No.1
ECCENTRIC Mild Steel No.1
BACK SHAFT AND NUT Mild Steel No.1
ECCENTRIC Mild Steel No.1
Case Hardened Out Side
BACK GEAR ECCENTRIC PINS Mild Steel No.2
CONE COUPLING CLAMP No.1 Bess. Steel Iron Nut C.Hd.
CATCH PLATE STUD No.2 Wrot Iron
TAIL BUSH ADJUSTING SCREWS Mild Steel No 2
FAST HEAD ADJUSTING SCREWS Mild Steel Chased No.4
FAST HEAD CLAMP STUDS No.2 Wrot Iron
BACK SHAFT ECCENTRIC KEYS Bessemer Steel No.1 to each
Scale of Sheet 1½ in to 1 Ft.
FRONT BEARING BUSH NO.1 Gun Metal
TAIL BEARING BUSH NO 1 SET In Three Parts Gun Metal
Machined All Over
COVER PLATE FOR BUSH Gun Metal

SHEET 19 contains details of the wrot iron and steel work for the Back Gear spindle, eccentrics, hand lever, and locking pins &c., and sundry other fittings for the Fast Head. Also details of the taper "neck bush" and the "tail bush" and collar, in gun metal or phosphor bronze. These latter figures can be drawn on the "pattern" sheet as before mentioned. The eccentrics should be a good "fit" on the Back Gear spindle ends, the head end one being keyed fast, while the tail end should be well fitted but not driven hard as it is necessary to have the barrel off occasionally for cleaning out the spindle hole.

All the nuts should be a good fit on good threads. The Fast Head clamp studs fit into the bosses under the head casting for securing the Head Casting to the Gantry by the clamp plates. They are to be screwed into the bosses as tightly as they can be got in, as they are per-manent fixtures.

Those parts of the Back Gear spindle, on which the Back Gear runs when in use, should be case hardened and ground.

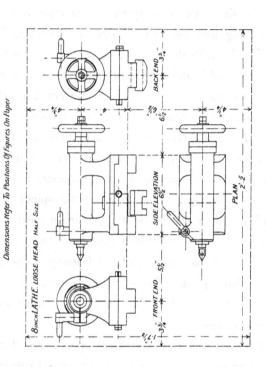

8″ LATHE, ARRANGEMENT of DRAWING of LOOSE HEAD　　SHEET 20

Dimensions Refer To Positions Of Figures On Paper

8 INCH LATHE. LOOSE HEAD HALF SIZE

BACK END　3¼″

SIDE ELEVATION　6½

PLAN　2′ 2″

FRONT END　5½″

3¾″

6½″

4¾″

1′ 7½″

SHEET 20 contains an arrangement of four views of the "Loose Head" of the lathe on a half double elephant sheet of drawing paper, as a general drawing of the head. This drawing is to be made half size or 6″ scale, and filled in with full detail from Sheets 21 and 22, reference to be made also to Sheets 23, 24 and 25 for dimensions and particulars.

The Head, in this example, consists of a block or base which fits and slides in the centre gap of the lathe bed or Gantry, on which is mounted (so that it can have a limited transverse movement) the Head proper. A clamp bolt passes through the combination, and by means of a clamp plate which engages the under inner edges of the lathe table or Gantry top, fixes the Head in any position, as may be desirable, to accommodate the length of the "work," or piece to be operated upon. The block or base contains a suitable screw passing from side to side and capable of being rotated by a "square" formed on one end. The screw has a collar at the square end and lock nuts at the other end which while admitting of rotation of the screw prevent any "end motion." The under side of the Head proper is fitted with a nut having a shank which engages in a hole in the Head proper; the nut embracing the screw mentioned and fitting well on the thread of the same. By these means the Head proper can be moved laterally for the purpose of turning tapers on "work," by setting the "centre" out of the centre line of the Gantry. The tool being fixed on the Saddle, which moves *parallel* to the centre line of the Gantry, produces taper "work," if the Fast Head and Loose Head centres are not parallel to the movement of the Saddle.

In setting the Heads for turning "parallel" work, that is, of uniform diameter, the Fast Head is first secured in position parallel to the centre line of the Gantry, by the setting screws and the clamp studs. The spindle is supposed to be mounted perfectly horizontal to the planed under surface of the Head. The Loose Head "centre" is then placed close to the Fast Head "centre" and the loose head is adjusted laterally, by the "taper block" screw mentioned, as near as can be by sight. The Heads are then moved apart and a perfectly parallel spindle (which is usually kept for the purpose) is placed between the centres (as for turning). By means of a dummy tool fixed in the tool holder of the saddle, trial is made, first at one end of the parallel spindle and then at the other, and repeated, as the lateral adjustment of the Loose Head is made by the "taper screw," until the two centres are seen to be dead "in line" with the movement of the Saddle on the Gantry.

Turned work with taper ends, as piston rods, tap plugs and so forth, can be made by suitable setting "out" the Loose Head "centre" by means of the taper block and screw.

8" LATHE, LOOSE HEAD DETAIL (Side and Plan)

SHEET 21

Scale 1½ in. = 1 Ft.

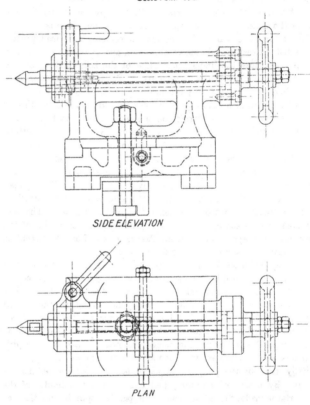

SIDE ELEVATION

PLAN

To Be Scaled Off or Drawn From Detailed Parts

SHEET 21 contains a Side Elevation and Plan of the Loose Head, to $1\frac{1}{2}''$ scale of the Book figure.. These are to be drawn into the General Drawing, half size, or four times the Book scale.

The arrangements for the Transverse motion of the Head proper are clearly shewn in both the views. The main feature of the Loose Head is the "Screw spindle" which consists of a cylindrical stout bar, which is projected from the head barrel by means of a screw which is held by a collar to the outer end of the barrel, and which passes through a nut which is solid with the screw spindle, and whose length is "housed" in a hole passing through the spindle. The "Screw spindle" is prevented from turning in its guide hole by a key which engages in a groove from end to end of the under side of the spindle. The term "spindle" is a misnomer, as the part is really a slide.

Both spindle and guide hole are very carefully machined and "lapped" and "ground" to a very close "fit." Spindle case-hardened. The screw shank has a square end which is fitted with a hand wheel for turning the screw. The movement of the screw spindle is mainly used for inserting the "centre" (which is similar to that of the Fast Head) in a "centre hole" which is made in the end of the bar or the like, which is to be turned, for supporting that end and providing a stable point of rotation.

For taking up any slack in the slide or "screw spindle" hole, one side of the barrel, at the inner end, is split, so that by means of a pair of lugs and clamp bolt the bore can be contracted and made to take a tight hold on the spindle end. The "centre," it will be seen, is inserted in a conical enlargement of the screw "housing" or sheath hole.

The clamp action on the barrel is brought about by operating the lever-nut.

For dimensions and particulars reference is to be made to Sheets 23, 24 and 25.

8" LATHE, LOOSE HEAD DETAIL (Front and Back)

SHEET 22

Scale 1½in = 1 Ft

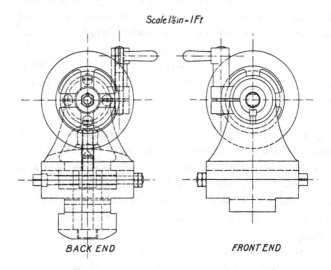

BACK END   FRONT END

To Be Scaled Off or Drawn From Detailed Parts

SHEET 22 contains the back and front End Elevations of the Loose Head. The back end elevation shews the taper-screw and nut very clearly and also the Head clamp plate, bolt and nut.

The front End Elevation shews the barrel clamp, stud and lever nut ; and the key which engages in the long groove on the under side of the Spindle slide. The key consists of a tee head on a round shank which drops into a hole in the under end of the barrel.

The " centre," like that of the Fast Head spindle, is provided with two flats to some spanner standard for twisting the centre loose, when it has to be withdrawn for any reason. Sometimes, too, the screw is made long enough to force the centre out of its socket, if it happens to " set fast."

These views are of course to be drawn into the general drawing arranged as shewn in Sheet 20.

8" LATHE, LOOSE HEAD CASTING DETAIL
(Side, Back End, Plan, Front End, Wheel)
SHEET 23

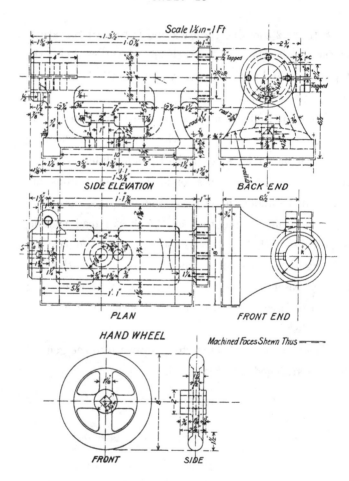

SHEET 23 contains detailed views of cast iron parts of the Loose Head; the Head proper, in Side Elevation, Plan and front and back End Elevations. Also the hand wheel.

These views, together with those on Sheet 24, can be drawn on one "pattern" sheet or drawing, which also serves for the machine shop.

The solid parts of the Head are "lightened out" by "cores," for the purpose, mainly, of maintaining something near uniform thicknesses of metal in the casting, and ensuring "soundness" by providing outlets for the gas generated by the molten metal.

All the machine work on the Loose Head casting should be carefully and exactly executed. The bore of the barrel should be perfectly parallel and exactly "square" with the lugs which embrace the projection on the base casting, and the bore should also be exactly parallel to the fitting surface of the base on the lathe Gantry. If the "centre" is at all out of " truth " with the centre line of the "work" it is constantly atwist in the conical centre hole in the "work," and its "bearing" value is destroyed.

8" LATHE, LOOSE HEAD BASE BLOCK DETAIL
(Side, End, Top, Under, Clamp, Plan)

SHEET 24

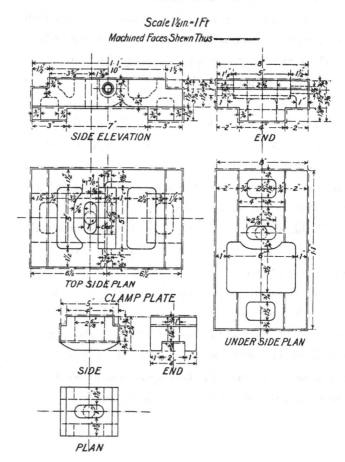

SHEET 24 contains Top side Plan and Under side Plan as well as Side Elevation and End Elevation of the base block of the Loose Head, also Side, Plan and End of the clamp plate.

The foregoing are also "lightened out" as far as possible, while maintaining the rigidity of the casting.

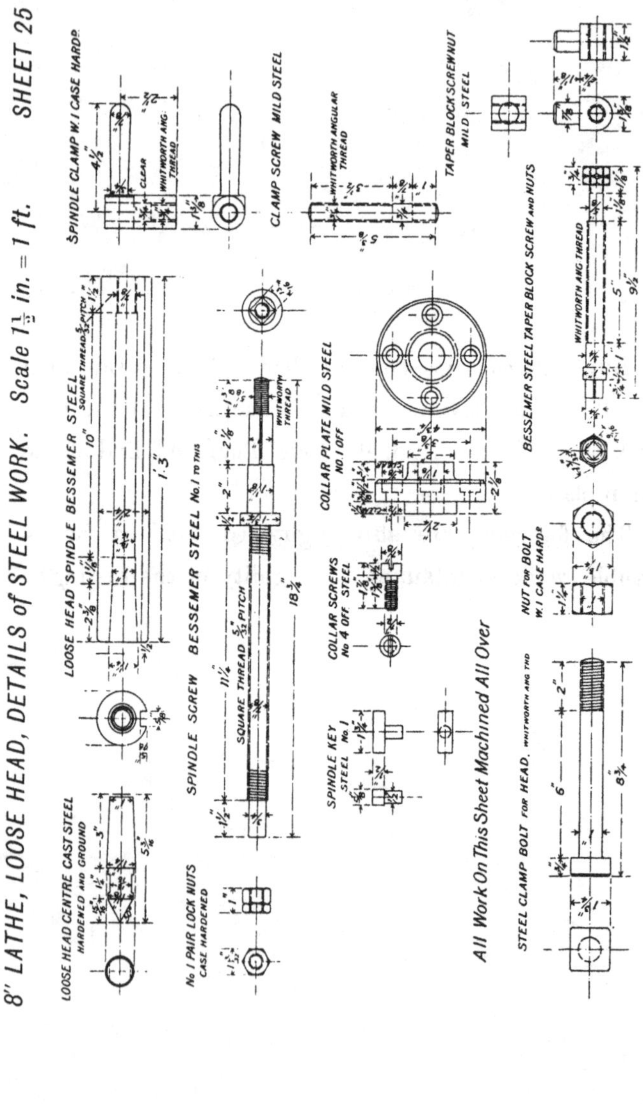

8" LATHE, LOOSE HEAD, DETAILS of STEEL WORK.   Scale 1½ in. = 1 ft.   SHEET 25

SPINDLE CLAMP W⁄I CASE HARDD.

CLAMP SCREW MILD STEEL

TAPER BLOCK SCREW NUT MILD STEEL

WHITWORTH ANG. THREAD

WHITWORTH ANG. THREAD

LOOSE HEAD SPINDLE BESSEMER STEEL
SQUARE THREAD

COLLAR PLATE MILD STEEL
NO I OFF

BESSEMER STEEL TAPER BLOCK SCREW and NUTS
WHITWORTH ANG THREAD

SPINDLE SCREW   BESSEMER STEEL No I TO THIS
SQUARE THREAD ⅝ PITCH

COLLAR SCREWS
No 4 OFF STEEL

NUT FOR BOLT
W⁄I CASE HARDD.

LOOSE HEAD CENTRE CAST STEEL
HARDENED AND GROUND

No I PAIR LOCK NUTS
CASE HARDENED

SPINDLE KEY
STEEL No I

All Work On This Sheet Machined All Over

STEEL CLAMP BOLT FOR HEAD. WHITWORTH ANG THD

SHEET 25 contains all the steel and iron parts of the Loose Head, detailed for use in the Smiths' Shop and Machine Shop. The whole of these parts are to be of the best class of material and work, and case hardened all over spindle slide, and on all bearings, squares, nuts and handles. The spindle slide especially should be a good job and well fitted in the bore of the head barrel.

This Sheet completes the Loose Head drawings.

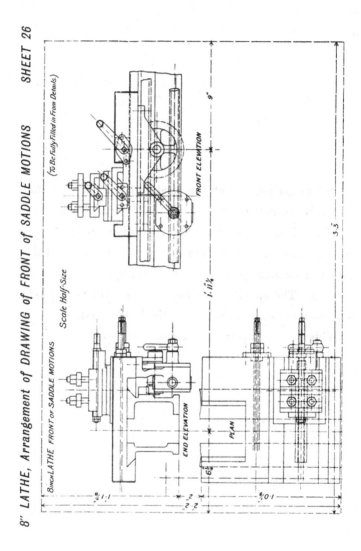

8" LATHE, Arrangement of DRAWING of FRONT of SADDLE MOTIONS   SHEET 26

(To Be Fully Filled in From Details)

Scale Half-Size

8 INCH LATHE FRONT OF SADDLE MOTIONS

FRONT ELEVATION

END ELEVATION

PLAN

SHEET 26 contains a model arrangement of drawing for shewing the mechanism which is connected with the working of the movements in the Front of the Lathe Saddle. The scale of this drawing is to be 6″ to a foot (half size) and the drawing is to be made on a double elephant sheet of paper (39″ × 26″ nett). This method of getting out the drawings of the Saddle and connections is adopted as corresponding in a measure to group design which is generally used in connection with machinery which contains several distinct sets of parts, which are connected only by the main drive; and also on account of the large size of paper on which a general drawing of a sufficiently large scale to properly admit the minutiae would have to be made, if both back and front of the Saddle were shewn on the same sheet. In a general drawing of a relatively small machine, in which all the parts are to be shewn, what is termed a "large" scale has to be used.

In accordance with this idea, Sheet 26 drawing deals with the mechanism which extends from the front of the Saddle to the centre line of the Lathe; and Sheet 30 in like manner deals with the mechanism at the back of the Saddle also up to the centre line of the Lathe. This admits of making the scale of the drawing half size, on a double elephant sheet of paper. Sheet 26 drawing is to be filled in with all the detail which is shewn (as pertaining to the group mentioned) on Sheets 27, 28 and 29; anything which is not sufficiently clear to the student on these sheets being contained in separate dimensioned detail on Sheets 32, 33, 34, 35, 36, 37 and 38, and reference should be made at once, in case of any doubt in the scaling of the parts, off the three sheets first mentioned (27, 28, 29).

At this stage it is assumed that the student has attained to considerable proficiency as a draughtsman, and will be well able to cope with the intricacies of such a drawing. The drawing is of a kind which is common in drawing office practice and which must be mastered.

The scale of the Book drawings in the sheets mentioned is $1\frac{1}{2}''$ to a foot with the exception of some on Sheet 35, which are half size. There will therefore be no particular difficulty in scaling the full views (27, 28, 29) and the others are fully dimensioned.

The general purpose of the Saddle is to carry the cutting tool, and to provide such motions for it as are necessary for doing the various kinds of work. These may be described as "sliding," or producing cylindrical or slightly conical surfaces; "surfacing," or producing *plane* surfaces at right angles to the lathe centre line; and "screw cutting," which consists of ploughing helical grooves in cylindrical surfaces, as in making a screw or "worm." A "worm" is generally understood to be a coarse screw to be used for working in the teeth of a wheel ("worm wheel"). Any combinations, within limits, of these motions may be used. The Saddle of course slides on the bed of the Lathe Gantry and is held in position by dovetails which are provided with adjustments for taking up "wear" or "slack." The "sliding" and "surfacing" actions are produced by motion which is obtained from the Back Shaft of the lathe.

For "sliding," motion is brought from the Back Shaft, through the thickness of the Saddle "table," by a hollow shaft, on which at the front of the Saddle a wheel (spur) is mounted, the teeth of which engage in a larger spur wheel (mounted on a bracket) which is keyed on the same short shaft or spindle as a "pinion" which works in gear with a "rack" which extends the length of the lathe (or of the saddle motion). By these means the "sliding" motion of the saddle is "self acting" or automatic. This motion is put in or out of gear by a conical clutch on the back end of the hollow shaft mentioned, and the clutch is operated by a rod and nut, the rod passing down the centre of the hollow shaft and the nut clamping against the outer "front" end of it. The rod pulls on the cone of the clutch and the hollow shaft is turned by the friction of the cone. "Clamping" or "loosing" the nut, by means of an ordinary spanner, puts the "sliding" motion "in" or "out" of gear.

For "surfacing," motion is brought to a Cross Slide, which is mounted on a dovetailed guide, which extends from front to back across the saddle, exactly at right angles to the centre line of the lathe Gantry. The motion of the "Cross Slide" is across the Saddle, and is produced by a screw which works against collars in the direction of its length, and is rotated as desired by a similar clutch to that of the "sliding" action, from the Back Shaft. The thread of the screw works in a nut which is fixed on the under side of the cross slide. A clamping action for the clutch is similar to that described for "sliding."

In both "sliding" and "surfacing" actions of the Saddle, the rate of progress of the tool bears some proportion to the rot tionaof the Lathe

Spindle, the "traverse" being from, say, $\frac{3}{32}$ of an inch to $\frac{1}{50}$th or so per revolution of the spindle according to the gear combinations from the "tail wheel" to the Saddle, and can be varied at short intervals of difference. But the "driving" of these actions, going through belt motions which may "slip" and always "creep," and friction clutches which may slip, are not sufficiently definite for making any but plain surfaces by tool points which are wider than the "traverse," and therefore overlap on the "cut."

For cutting or ploughing spiral grooves to a definite pitch with perfect accuracy, as in making a screw or worm, wheel gear is used for actuating or turning a large screw termed the Leading Screw (which extends the length of the lathe), which engages definitely with a nut which is fixed to the Saddle. The Saddle thus has a "coupled up" definite rate of progress, imparted from the Spindle "tail wheel," which has no liability to change from any "slipping" of connections like the other actions, and which therefore will cut a "thread" or the like with an accuracy which is equal to that of the Leading Screw and driving gear. The Screw Cutting action being "coupled up" some provision has to be made for starting and stopping the "cutting" action, as required by the work, and this is done by "splitting" the Nut (making it in two halves) and providing a "cam" action for "opening" and "closing" the Nut on the threads of the Leading Screw. In the "open" position, the halves of the Nut are quite clear of the screw, and in the "closed" position they are in correct mechanical engagement with it. For manipulating the Nut, a fitting termed the Nut Box is provided and fixed to the underside of the Saddle in such a manner that the screw passes through the two halves of the Nut, the halves working in proper guides, and having "cam" slots cut through them, by means of which two pins, mounted on a disc having a spindle and lever, "open" and "close" the Nut as required. When the nut halves are in gear they are "locked" by a concentric portion of the "cam" slots, and when they are out of gear they can be "clamped" in a safe position by the nut which secures the lever on the pin disc spindle. These provisions will be recognised on Sheet 27.

Clamps are provided to all the "slides" including the Saddle itself, for fixing firmly and free from "shake" every part but that which is actually "working."

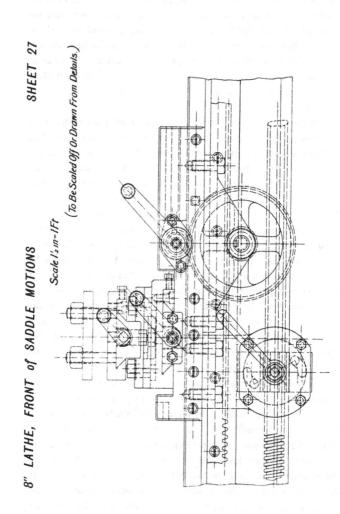

8" LATHE, FRONT of SADDLE MOTIONS　　SHEET 27

Scale 1½ in=1Ft

(To Be Scaled Off Or Drawn From Details.)

SHEET 27 contains a front Elevation of the Saddle in which the Rack and wheel and pinion which engage with it are plainly shewn. The wheel is mounted in a bracket which is screwed on, under the edge of the Saddle, a key securing the wheel boss on a short shaft, which has a "square" end on which the pinion is mounted. The wheel is driven by a smaller wheel which is fixed on a "square" on a tubular shaft which passes through the thickness of the Saddle from back to front and which obtains its motion from the back shaft by means of a taper clutch. This tubular shaft passes through a flanged collar which keeps it in place, and at its front end has a "square" on which a loose handle can be placed. When the clutch is disengaged, the traversing motion can be produced by the handle, which by means of the small wheel, spur wheel and pinion, and rack can move the Saddle in either direction endwise of the Saddle. A rod passes through the tubular shaft, from the clutch, and comes to the front end, where it is provided with a nut which can be tightened against the tubular shaft end, thus putting the rod, and consequently the taper clutch into gear; this couples up the traversing motion which is received from the Back Shaft of the lathe, and makes the "traverse" self acting.

The "Leading Screw" and split Nut (made in halves) are also plainly shewn, as well as the pins and cam slots by which the opening and closing actions of the Nut are brought about. The construction of the pin disc and the lever by which it is worked are shewn separately on Sheet 36; and the split Nut and cam slots on Sheet 34. The concentric portion of the slots locks the Nut in the closed position. When the hand lever is "down," the Nut is closed, and when it is "up" the Nut is open and out of gear with the leading screw thread. In this "open" position the Nut can be "locked" by tightening the lever nut outside the lever boss, which clamps the whole arrangement and prevents the Nut from dropping into gear accidentally or inadvertently: an important precaution. On Sheet 28, the construction and fitting of the "Nut box" is shewn; as also the wheel and pinion in gear with the rack.

The Saddle casting has a large dovetailed guide right across from front to back, on which a slide (provided with adjustments) is mounted, which by means of a screw, which receives motion by a taper clutch from the back shaft, can be "traversed" the length of the dovetailed guide, "self acting"; or by hand, by means of a handle (or sweep) which can be used on a "square" on the front end of the screw. The slide is termed the "cross slide" of the Saddle. On the upper side of the "cross slide" casting a turn-table is formed, and on the turn-table another dovetailed guide with adjustable slide is mounted. This combination is termed the "slide rest"; on the upper table of this the actual tool holder is arranged, consisting of four studs and two cross bars. The cutting tool is placed under the cross bars and the nuts tightened down so as to fix the tool firmly in the position desired.

Sheets 28 and 29 shew the parts indicated in the Saddle, and Sheets 35 and 36 shew the screw, shaft and fittings separately.

Sheet 32 provides particulars of the Saddle casting, fully prepared for receiving all the fittings.

This drawing is to be transferred to the general drawing, Sheet 26, and not necessarily made separately.

The Book scale is $1\frac{1}{2}''$ to a foot, and the figure can be measured by the scale, but checking of sizes can be done by referring to Sheets 28, 29, 32, 33, 34, 35, 36, 37 and 38.

8″ LATHE, CROSS SECTIONS of FRONT and BACK of SADDLE MOTIONS　　　SHEET 28

To Be Drawn Half-Size To Sheets 26 And 30.

Scale 1½in.=1Ft.

(To Be Scaled Off Or Drawn From Detail)

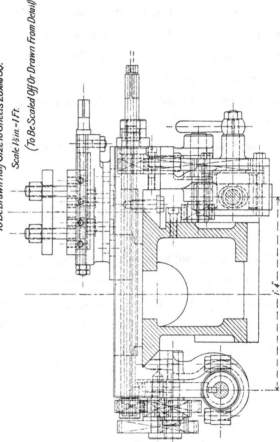

SHEET 28 contains a Cross Section of the lathe Gantry, and End Elevation of the Saddle with full fittings, including the Cross Slide and Slide Rest.

The Leading Screw and Back Shaft are in section. The Leading Screw is seen to pass through the "Nut Box," clear of the hole in the box and of the Nut, when that is "open" as shewn. Part of the Nut "thread" is cut away, so that what remains can lift or fall clear of the Leading Screw without excessive movement. The "cam" pins and disc and lever can be clamped, in either the "open" or "closed" position of the Nut, by means of the spindle end nut. In the "closed" position of the Nut (Leading Screw Nut) the cam-slots lock the Nut in position and the use of the clamp nut is not usually necessary; but in the "open" position of the Nut it is important that it should be secured from accidentally falling into gear with the Leading Screw, and the use of the clamp is advisable.

The Back Shaft passes through the boss of a worm, which it turns by means of a long groove, which engages a "sliding key" provided in the worm boss. The worm works a worm wheel which in turn actuates properly graded gear wheels, which operate the "clutch wheels" on the ends of the Cross Slide "traversing screw" and the "hollow shaft" of the "sliding" action.

The Leading Screw half of the Saddle detail shewn in this Sheet is to be drawn into the General Drawing outlined in Sheet 26.

The Back Shaft half goes into Sheet 30.

The Cross Slide, turn-table and Slide Rest, go into Sheet 26.

The centre lines of the dotted wheels and parts, and dimensioned particulars are to be obtained from the Saddle Plan (Sheet 29) and Sheets 33, 34, 35, 36, 37 and 38.

The half size drawing (to Sheet 26) will give space for shewing the dotted details with sufficient clearness.

Sheet 31 provides the centres of the gear studs and particulars of the positions of the parts relating to the Back Shaft.

The Scale of this Book figure is $1\frac{1}{2}''$ to 1 foot.

Some patience is required in dealing with the dotted work, and it is facilitated to a considerable extent by drawing in the solid line parts first, so that the extent of the dotted lines may be readily seen; and by maintaining uniformity in the strength of the lines and avoiding any rubbing of the paper.

By this time the young draughtsman will have acquired considerable proficiency in his work and will be able to execute complicated drawing. Moreover as this kind of work comes into the regular practice of a drawing office, it has to be done in the course of instruction.

Repetitions in these explanations are made to save back reference and confusion.

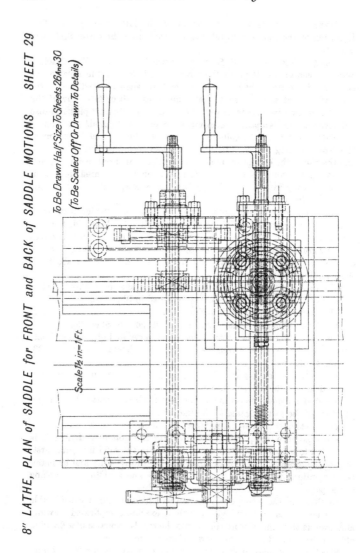

8" LATHE, PLAN of SADDLE for FRONT and BACK of SADDLE MOTIONS   SHEET 29

To Be Drawn Half Size To Sheets 26 And 30

(To Be Scaled Off Or Drawn To Details)

Scale 1½ in=1 Ft.

SHEET 29 contains the Saddle Plan, on which there is some close dotted work.

The right-hand half is concerned with the General Drawing on Sheet 26, and the left-hand half with the drawing to Sheet 30.

The Saddle casting fully prepared for all the parts is found on Sheet 32 and will be useful in locating the fixings, screws, &c. The "coreing" or "lightening out" of the casting will be noted. This is an important detail.

The Cross Slide screw, clutch and fittings are shewn on Sheet 37 and the Cross Slide and Slide Rest castings are on Sheet 40; while the fittings for the Slide Rest and the hollow shaft and fittings are on Sheet 38. The rack pinion and shaft, the Back Shaft gear studs, tool clamp bars and studs, handles, sundry screws &c. are on Sheet 39.

The Leading Screw and Nut Box are omitted, to save undue and useless complication under the Slide Rest.

The hollow shaft through the Saddle which brings the motion to the "rack traverse" for sliding, and the screw which brings the motion to the Cross Slide, are both readily seen, as also the hand motions to each. The positions of the rack, pinion, spur wheel and driving wheel on the hollow shaft are quite clear.

The dotted work will be put in most readily by first drawing in the bracket castings, Sheet 33, and then filling in the parts belonging to these from Sheet 31 and Sheet 37.

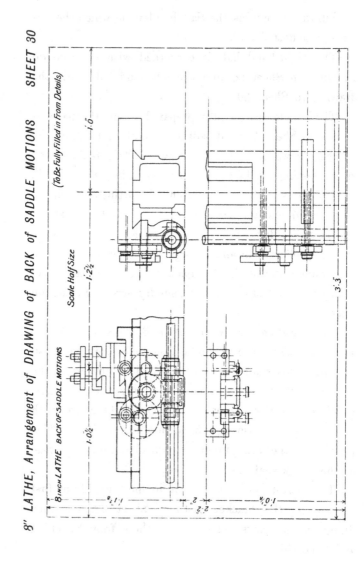

8" LATHE, Arrangement of DRAWING of BACK of SADDLE MOTIONS   SHEET 30

(To Be Fully Filled in From Details)

Scale Half Size

8 INCH LATHE   BACK OF SADDLE MOTIONS

SHEET 30 contains a model arrangement of drawing for shewing the arrangement and working of the transmitting mechanism from the Back Shaft to the hollow shaft and screw of the sliding and surfacing motions.

The drawing is to be made half size on a double elephant sheet of paper, the objects of it being similar to those of Sheet 26.

Full detail is to be put into the drawing from Sheets 31, 32, 33, 35, 36, 37 and 38, as in the former case also.

The Back Shaft, sliding key in worm boss, worm bracket, worm wheel and the other gear are plainly shewn in Sheet 31.

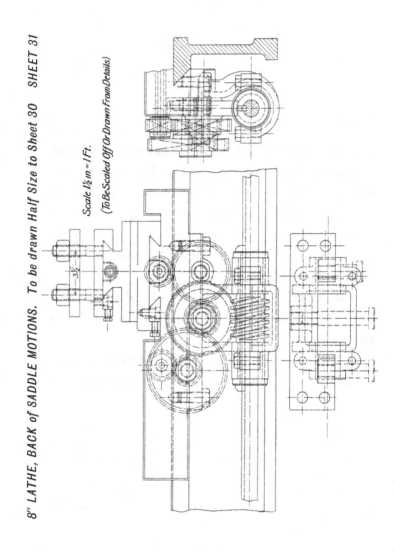

8″ LATHE, BACK of SADDLE MOTIONS.   To be drawn Half Size to Sheet 30   SHEET 31

Scale 1½ in. = 1 Ft.

(To Be Scaled Off Or Drawn From Details)

SHEET 31 contains a fully detailed drawing of the arrangement of mechanism for the back of the Saddle, for taking the driving actions of the Sliding and Surfacing motions from the Back Shaft. The B. S. is seen to pass through the bearings and boss of a worm (Sheet 33) which it revolves by means of a sliding key which is carried by the worm. The worm is mounted in a pair of bearings which are formed in a bracket which is fixed under the edge of the Saddle by means of four screws. The worm gears into a worm wheel, to the boss of which two separate pinions are fixed, and the whole worm wheel and pinions revolve on a stud which is fixed into the bracket. These pinions impart motion respectively to trains of wheels connecting the hollow shaft and screw of the sliding and surfacing actions. The wheels and screw are made so as to give equal "traverses" to the sliding and surfacing motions. A plan of the bracket is shewn to make the parts as clear as possible in the drawing. The contents of this sheet are to be drawn into the general drawing, Sheet 30, and not necessarily drawn separately. The Scale of the Book figure is $1\frac{1}{2}''$ to a foot, and measuring off by the scale will not be difficult; but as in former cases reference can be made to the dimensioned details on Sheets 32, 33, 34, 35, 36 and 37.

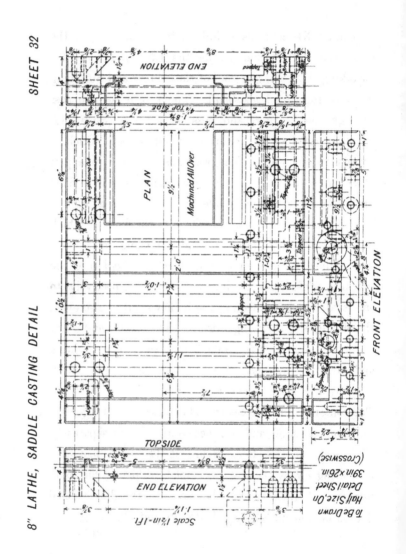

SHEET 32

8" LATHE, SADDLE CASTING DETAIL

PLAN

END ELEVATION

TOP SIDE

FRONT ELEVATION

END ELEVATION

TOP SIDE

To Be Drawn
Half Size, On
Detail Sheet
39in x 26in.
(Crosswise)

Scale 1½in.=1 Ft.

SHEET 32 contains a drawing in three views, of the lathe Saddle casting, fully dimensioned and particularized for the Pattern Shop or Machine Shop. It can be regarded as a good general guide for making all the drawings in which the Saddle is concerned.

This drawing is to be made half size on a half double elephant sheet of paper or on the space of half such a sheet; the remainder may be filled up by the details of Saddle fitting castings on Sheets 33 and 34, all of which come under the same heading as to class of work. The book scale is $1\frac{1}{2}''$ inches to a foot, and the figure with dimensions is sufficiently clear to be easily understood and measured. The casting is hollowed or "lightened out" to avoid very thick places which might be "spongy" or unsound if left solid. Such lightening will be recognised. This Sheet is very important and should be carefully noted throughout.

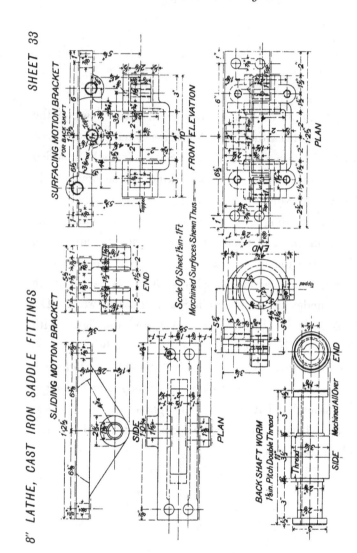

SHEET 33

8" LATHE, CAST IRON SADDLE FITTINGS

SURFACING MOTION BRACKET
FOR BACK SHAFT

FRONT ELEVATION

PLAN

END

Scale Of Sheet 1½in=1Ft
Machined Surfaces Shewn Thus

SLIDING MOTION BRACKET

SIDE

PLAN

END

BACK SHAFT WORM
⅛in. Pitch Double Thread

SIDE   Machined All Over   END

SHEET 33 contains detail drawings of the front and back brackets of the Saddle which carry all the parts for the motions, except the Nut Box.

These should all be carefully noted, as they form good guides for correct drawing of the parts in the general drawings. The worm "blank" (uncut) only is shewn, but finished sizes as to the other features.

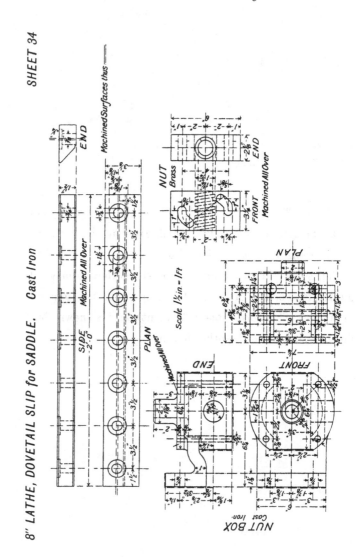

SHEET 34

8" LATHE, DOVETAIL SLIP for SADDLE. Cast Iron

SHEET 34 contains drawings of the "dovetail slip" or wear adjustment of the Saddle, which is also used for clamping the Saddle in position when the "surfacing" motion is in action. Also the Nut Box and Nut are shewn. The construction of these is self evident. The Nut is machined all over, as also is the inside of the "box" guide in which it works. The outside lid which carries the cam pin spindle boss or sleeve, is also machined all over. The "leading screw" hole, through the box, is also of course bored true in position, and forms a support for the screw and prevents it "sagging." The joint face where the box casting is fixed to the under edge of the Gantry is also correctly machined and drilled.

The Saddle "dovetail slip" is machined all over and scraped to a face with the dovetail flange of the "top table" of the Gantry. The two end screw holes, at the least, are to be a "fit" on their screws, at the sides, to act as "steady pins" endwise. These holes are all seen to be "slotted" to allow of movement for adjusting the "slip" close to the Gantry dovetail.

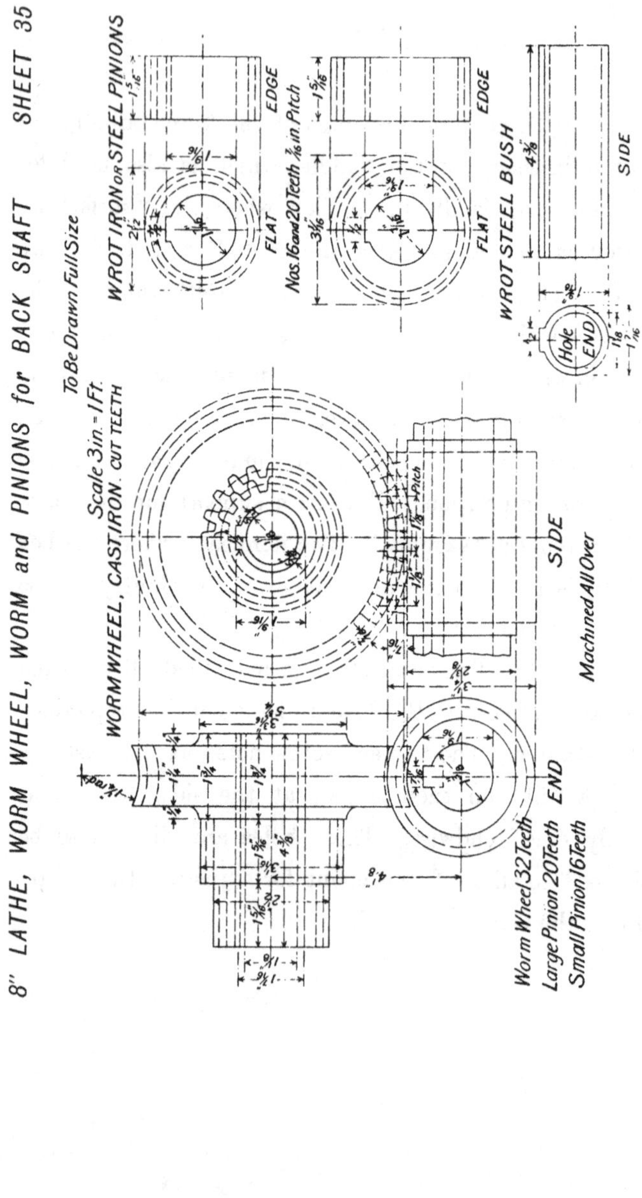

8" LATHE, WORM WHEEL, WORM and PINIONS for BACK SHAFT   SHEET 35

To Be Drawn Full Size

Scale 3 in. = 1 Ft.

WROT IRON or STEEL PINIONS

EDGE

FLAT

Nos. 16 and 20 Teeth ⅞ in. Pitch

EDGE

FLAT

WROT STEEL BUSH

SIDE

Hole

END

WORM WHEEL, CAST IRON. CUT TEETH

Pitch

SIDE

Machined All Over

END

Worm Wheel 32 Teeth
Large Pinion 20 Teeth
Small Pinion 16 Teeth

SHEET 35 contains a drawing in two views of the worm wheel and its two pinions and the worm. The worm wheel and the pinions are al "threaded" on to a key bush, which thus connects the three, the smaller pinion being too small in diameter to be mounted on any cast iron boss projecting from the wheel as both boss and wheel would be too thin for practical construction. The key bush can be made either of gun metal or mild steel. The whole three wheels are driven on, fairly tight. The top, bottom and pitch lines of the teeth are shewn of all three wheels, and in the Front Elevation the teeth of the pinions are drawn, and the teeth of the worm wheel and the thread of the worm are dotted for the sections on the centre line of the worm and wheel as shewn in the Side Elevation. These are for making templates of the worm grooves. The worm thread is *double*, the pitch being $1\frac{1}{8}''$ on the pitch line, while the thread and space is $\frac{9}{16}''$. The worm wheel is "cut" by means of a "hob," which is a milling cutter of facsimile form to that of the worm. The shape of the teeth is formed by the form of the worm thread and therefore accurate for that form. The worm teeth are first "nicked in" all round by a straight ring cutter, to give the "hob" a hold and start. The "hob" is then gradually "run in," as both "hob" and wheel revolve as in working. Separate "details" of the pinions and the key bush are shewn to make the arrangement quite clear. The pinions are both to be of either wrot iron or mild steel. The worm wheel is of cast iron, as also is the worm.

This Sheet is to be drawn full size on a double elephant sheet of paper; the teeth to be carefully set out, to shew the form, but not all round the wheels; only a few to each size, for making the "cutter blanks."

The Book figures are $3''$ to 1 foot.

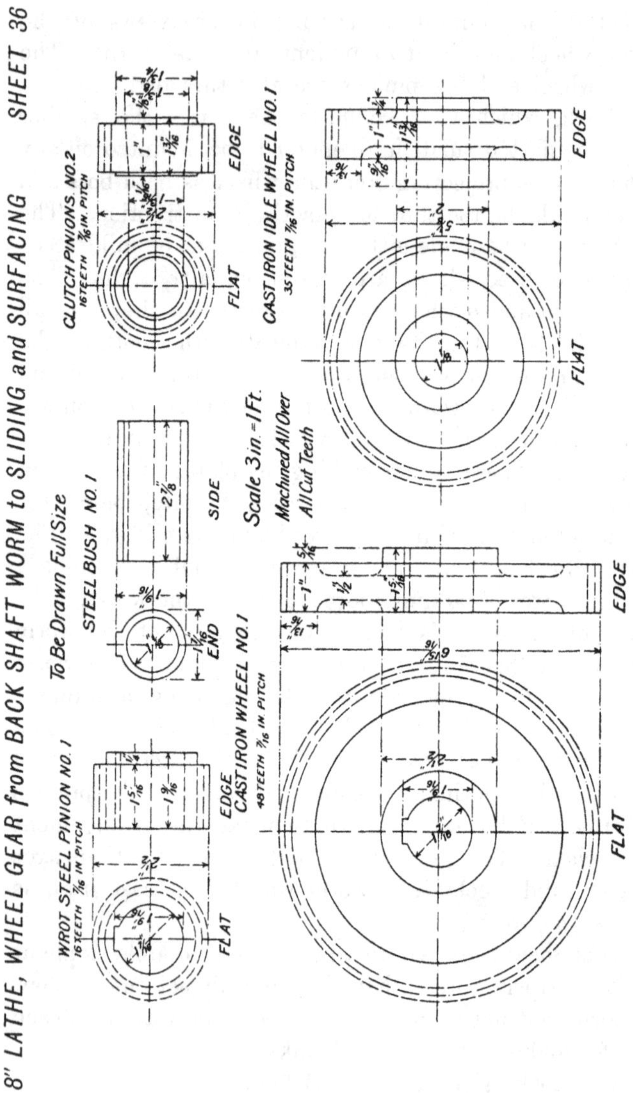
SHEET 36
8" LATHE, WHEEL GEAR from BACK SHAFT WORM to SLIDING and SURFACING
To Be Drawn Full Size
WROT STEEL PINION NO. 1
16 TEETH 7/16 IN PITCH
EDGE
FLAT
STEEL BUSH NO. 1
END
SIDE
CLUTCH PINIONS NO. 2
16 TEETH 7/16 IN. PITCH
EDGE
FLAT
Scale 3in. = 1 Ft.
Machined All Over
All Cut Teeth
CAST IRON WHEEL NO. 1
48 TEETH 7/16 IN. PITCH
EDGE
FLAT
CAST IRON IDLE WHEEL NO. 1
35 TEETH 7/16 IN. PITCH
EDGE
FLAT

SHEET 36 contains the cast iron wheel, wrot iron or steel pinion and key bush for reducing the motion between the smaller of the worm wheel pinions and the hollow shaft clutch wheel, of the "sliding" motion. It also contains the mild steel clutch wheel with taper bore, and the cast iron "idle wheel" which carries the motion from the larger worm wheel pinion to the screw clutch wheel, for the "surfacing" action. All these wheels have machined cut teeth and are otherwise machined all over.

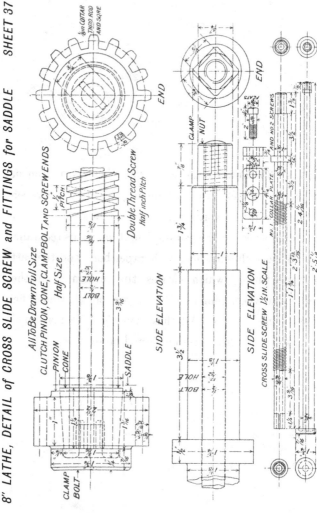

8" LATHE, DETAIL of CROSS SLIDE SCREW and FITTINGS for SADDLE   SHEET 37

SHEET 37 contains a half size "detail" of the ends of
the "cross slide screw" of the Saddle, and also a full
length "detail" of the complete screw and fittings to $1\frac{1}{2}''$
scale.

The screw end, communicating with the Back Shaft
motion, is formed into a bearing, and further has a
"square" for receiving the clutch cone, both of which are
clearly shewn. The "clutch wheel" fits on to the cone,
and is "pulled on" to make sufficient friction to "carry"
the screw without slip, by the head and long shank and
nut of the clamp or clutch bolt, which passes down the
centre of the screw barrel, and emerges on the end of the
screw "handle square." The clutch bolt nut is seen to be
square and of such a size that the square hole of the
handle will pass over it. By tightening the bolt nut,
against the end of the screw "square," the clutch is
brought into action and the screw driven by the Back
Shaft. When the nut is "slacked" the "clutch wheel"
runs loose on the cone.

The screw is $\frac{1}{2}''$ pitch, double thread; that is, there
are two threads, the revolutionary pitch or progress of
which is half an inch.

The "traverse" of the *cross slide* of the Saddle is made,
by the proportions of the "gear," to be the same as the
"sliding" traverse of the Saddle.

All the work on the parts shewn in this Sheet is of
course machined all over, and of the best class.

This Sheet is to be part of a sheet of steel or iron
"details" such as is required for the Smith's and Machine
Shops. Sheets 38 and 39 are to be incorporated in one
drawing with this Sheet. All the articles being relatively
small would be drawn full size. The spaces between the
figures for dimensions etc., would be about twice the size
shewn in the Book.

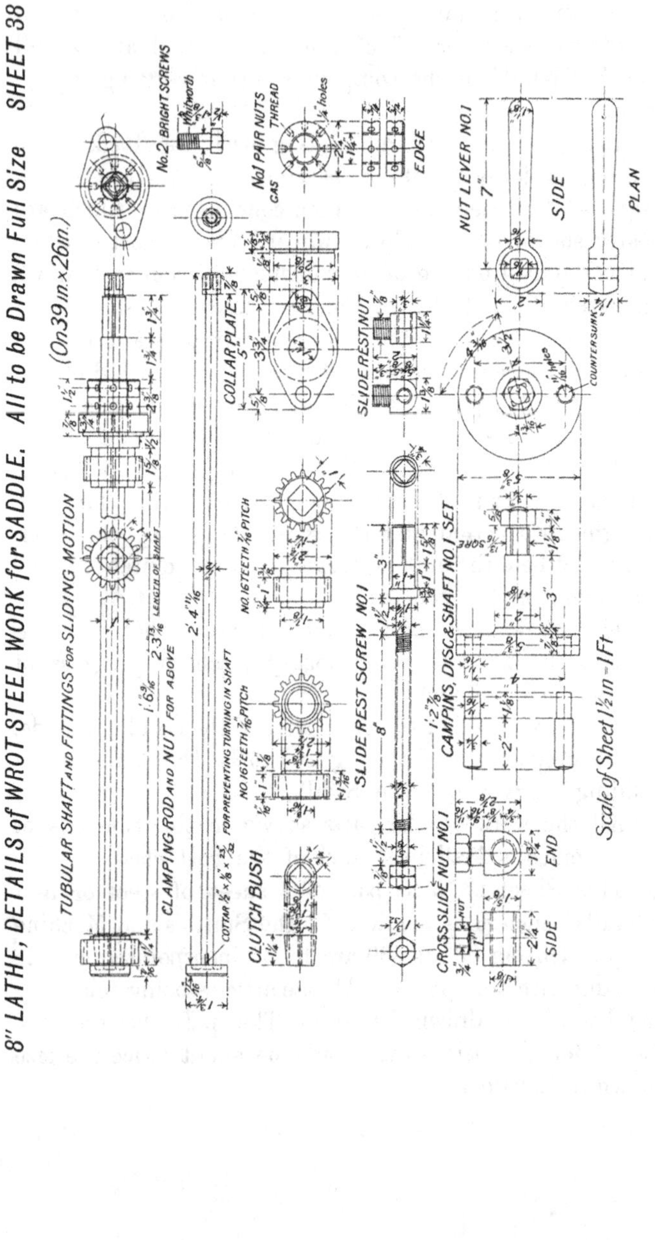

8" LATHE, DETAILS of WROT STEEL WORK for SADDLE.   All to be Drawn Full Size   SHEET 38

(On 39 in. x 26 in.)

TUBULAR SHAFT and FITTINGS for SLIDING MOTION

No. 2 BRIGHT SCREWS

No. 1 PAIR NUTS
GAS THREAD

NUT LEVER No. 1

COLLAR PLATE

SLIDE REST NUT

CLAMPING ROD and NUT for ABOVE

CLUTCH BUSH

FOR PREVENTING TURNING IN SHAFT
No. 16 TEETH ⅞" PITCH

No. 16 TEETH ⅞" PITCH

SLIDE REST SCREW No. 1

CAM PINS, DISC & SHAFT No. 1 SET

CROSS SLIDE NUT No. 1

Scale of Sheet 1½ in = 1 Ft

SHEET 38 contains "details" of the hollow shaft of the sliding motion, which passes through the thickness of the Saddle above the Gantry, for bringing the Back Shaft motion to the Saddle front gear.

The Scale of the Book figures is $1\frac{1}{2}$ inch to a foot, but all these parts being small they are to be drawn full size on a detail sheet as mentioned for Sheet 37.

This Sheet also contains separate figures of the "fittings" of the hollow shaft. It also contains details of the Slide Rest screw and fittings and details of the "cam pin" disc, spindle, nut and handle, for the Nut Box, all of which are also to be drawn full size on the shop detail sheet mentioned. All the parts are measurable by scale, but reference may be made to Sheet 37 for particulars of the "clutch wheel."

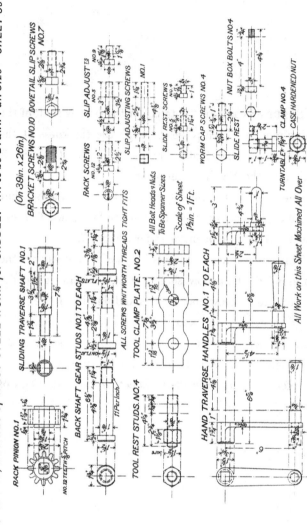

8" LATHE, DETAILS of WROT STEEL WORK for SADDLE.   All to be Drawn Full Size   SHEET 39

SHEET 39 contains "details" of the "traversing pinion" of the "sliding" motion, with shaft for the same and screws for the bracket. It also contains studs for the Back Shaft worm wheel gear, studs and clamp plates for the "tool rest" and handles for the squares of the Hollow Shaft, the Cross Slide screw and the Slide Rest. Also bolts and screws as enumerated.

All these are to go on to the detail sheet of drawings of the Smith Shop steel work. Speaking generally all the screws would be grouped together so that the blue print could be cut up; the screws being a separate item in the manufacture. As all the Sheets and Figures of the Book, however, have for their primary object the making of the drawings, strict shop methods of arrangement of details on the drawings are not here always adhered to; such as are most clear and convenient for reference, are sometimes preferred.

These details complete the steel and wrot iron parts of the Saddle.

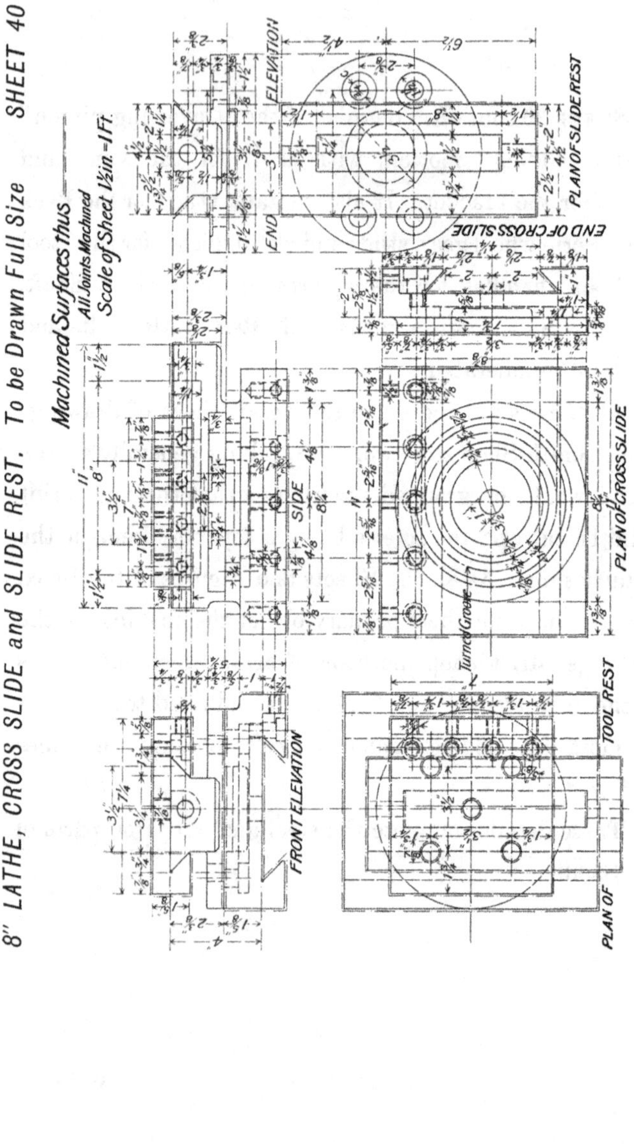

8" LATHE, CROSS SLIDE and SLIDE REST.　To be Drawn Full Size　SHEET 40

Machined Surfaces thus ——
All Joints Machined
Scale of Sheet 1½in.=1Ft.

ELEVATION

END

PLAN OF SLIDE REST

END OF CROSS SLIDE

SIDE

PLAN OF CROSS SLIDE

Turned Groove

FRONT ELEVATION

TOOL REST

PLAN OF

SHEET 40 contains the cast iron "details" of the Saddle Cross Slide and Slide Rest; with all the "work" of planing, turning and drilling on them shewn finished sizes. These are to be drawn half size and allowed more space between the figures on a double elephant sheet.

The views are a Front Elevation, Side Elevation and Plan of the castings of the Cross Slide Turn-table and Slide Rest in position, and the Cross Slide with Turn-table face in plan and end elevation; also a Front Elevation and Plan of the Slide Rest proper.

The Slide Rest is secured and clamped in any position by four T-headed bolts which engage in the undercut circular groove on the Cross Slide, and pass through the flanges of the Slide Rest slide casting.

The Cross Slide embraces the dovetailed guide across the Saddle by means of a fixed and an adjustable dovetail. The adjustable dovetail or "slip" is fixed by five cheese headed screws and "set up" by set screws.

These parts complete the Saddle drawings.

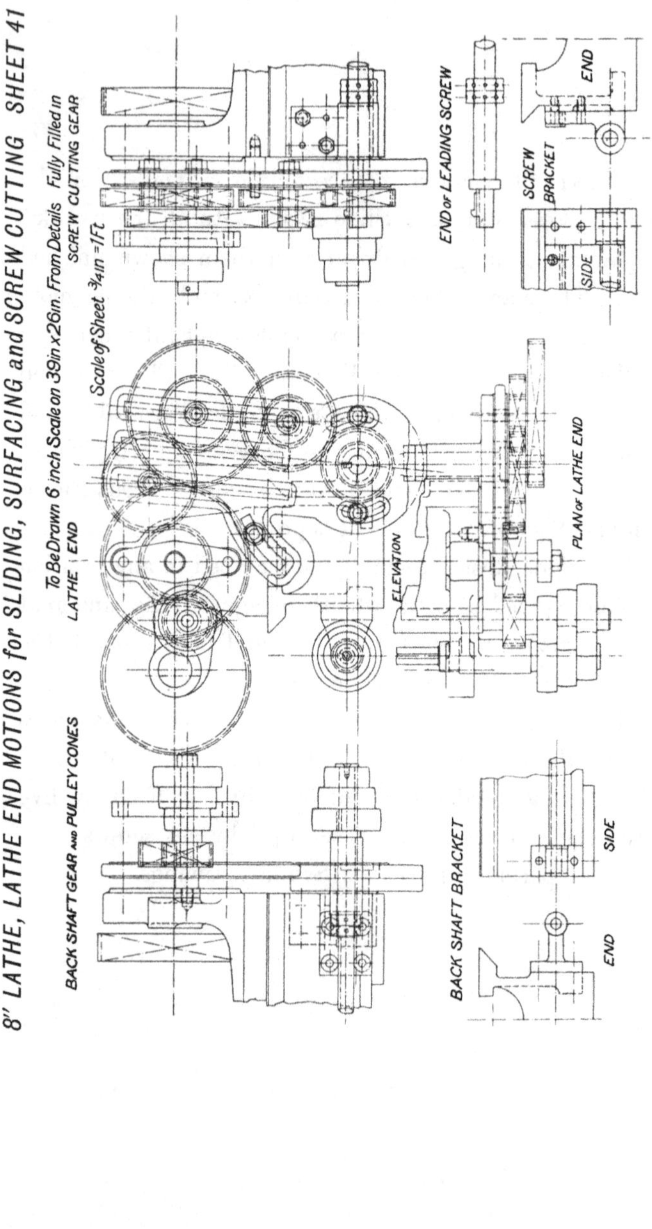

8" LATHE, LATHE END MOTIONS for SLIDING, SURFACING and SCREW CUTTING SHEET 41

To Be Drawn 6 inch Scale on 39in x 26in. From Details   Fully Filled in
LATHE ¦ END   SCREW CUTTING GEAR

Scale of Sheet ¾in =1Ft

BACK SHAFT GEAR ᴀɴᴅ PULLEY CONES

ELEVATION

PLAN ᴏꜰ LATHE END

END ᴏꜰ LEADING SCREW

SCREW
BRACKET

END

SIDE

BACK SHAFT BRACKET

SIDE

END

SHEET 41 contains a series of figures which shew the End Motions of the Lathe; which are such as convey the motion of the Tail Wheel of the Lathe and Spindle to the Back Shaft and to the Leading Screw. A "Motion," in these connections, signifies a series of parts, which convert and carry motion and power properly speeded from the source of power to the working parts proper.

In the End View of the Lathe, the Tail Wheel of the Spindle is seen to be connected with the Leading Screw by a series of wheel gears, which are termed "change wheels," and which are "set up" with the proper "counts" for giving any desired definite " traverse" for the cutting tool, in screw cutting. These are situated towards the front of the Lathe, or where the workman stands. Also in this view the same Tail Wheel is seen to be geared with a wheel on the boss of a "stepped pulley" or "cone," which by means of a short belt drives the back shaft "cone" and of course the shaft. The "steps" of the cones are in three pairs, giving three speeds, and the "wheels" from the Spindle to the stud cone (the upper one), can be changed to give two more speeds also, making 5 speeds which may be used for the Back Shaft actions of "Sliding" and "Surfacing." The "traverses" may be varied from $\frac{3}{32}''$ to $\frac{1}{54}''$. They are $\frac{3}{32}$, $\frac{1}{18}$, $\frac{1}{24}$, $\frac{1}{38}$, $\frac{1}{54}$ of an inch respectively.

This Sheet is $\frac{3}{4}''$ scale in the Book figures, and is to be drawn half size on a double elephant sheet; reference to be made to Sheets 42, 44, 45, 46, 47, 48, 49, 53, 54, and 55 for dimensions and particulars of the items.

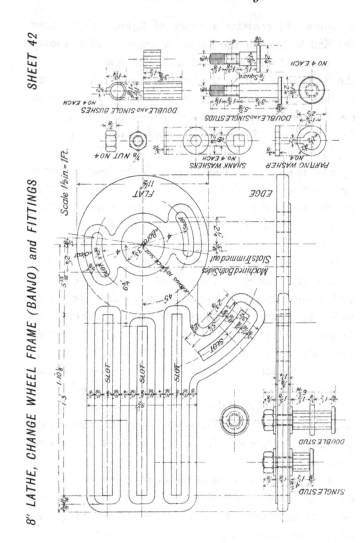

SHEET 42

8" LATHE, CHANGE WHEEL FRAME (BANJO) and FITTINGS

Scale 1½ in = 1 Ft.

SHEET 42 contains a detail drawing of the "change wheel frame" or "Banjo," of the End Motion together with the studs and key bushes for mounting the "change wheels." Also washers for nut, stud and separating the wheels.

The Scale of the Book figure is $1\frac{1}{2}''$ to a foot and the drawing is to be made $3''$ to 1 foot on a double elephant sheet, together with the figures on Sheets 44, 47 and 49. They are to be all arranged in the same way as shewn so that prints can be cut up.

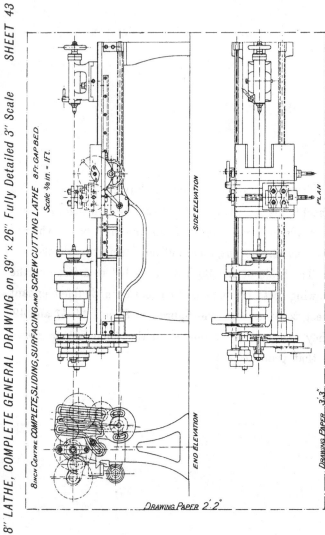

SHEET 43

8" LATHE, COMPLETE GENERAL DRAWING on 39" × 26" Fully Detailed 3" Scale

8 INCH CENTRE COMPLETE, SLIDING, SURFACING AND SCREW CUTTING LATHE 8 FT. GAP BED.

Scale 3/8 in. = 1 Ft.

SIDE ELEVATION

PLAN

END ELEVATION

DRAWING PAPER 2'.2"

DRAWING PAPER 3'.3"

SHEET 43 contains three views in outline of the complete Lathe. These form a model arrangement for a drawing on a double elephant sheet, to a scale of 3 inches to 1 foot.

The parts are to be drawn in from and as shewn on the different groups preceding, as the Gantry or Bed, Sheet 45; the Fast Head, Sheets 10, 11, and 12; the Front of the Saddle and the Plan of the Saddle, Sheets 27, 28 and 29; the Loose Head, Sheets 21, 22, 23, 24 and 25; the Back of the Saddle, Sheets 28, 29 and 31; the End Motion, Sheets 41, 42, 44, 46, 47, 48 and 49. Also from the general drawings of the groups and of all the other details shewn in the series including Change and other wheels. The whole series of 8″ Lathe Drawings from Sheet 10 to Sheet 55 are embodied in this General Drawing.

The model drawing is shewn to have an 8 ft. bed, which with the end motion is 9 ft. over all in length to scale, or 27 inches actual on the Drawing; the End Elevation is 2 ft. 10 in. over all, or $8\frac{1}{2}$ inches actual. The double elephant paper will be 39 in. long net, and the two items given amount to $35\frac{1}{2}$ in., leaving $3\frac{1}{2}$ in. for centre and end spaces.

The Book drawing is $\frac{3}{8}$ in. to a foot; the measurements are of course to be taken from the general drawings of groups and from the separate details, the Book figure simply shewing the arrangement.

In the Side Elevation the End Motion to Sheet 41 and details are to be filled in fully from the Leading Screw; the Back Shaft cone etc., in outline. The Fast Head is to be filled in fully from Sheet 10. The Leading Screw Brackets are to be filled in from Sheet 44. The Nut Box and Front of the Saddle is to be filled in from Sheet 28. The Loose Head is to be filled in from Sheet 21.

In the End Elevation the End Motion is to be fully filled in from Sheet 41 and details appertaining thereto. The Fast Head, Gantry, and part of Saddle is to be put in outline.

In the Plan the Fast Head with back gear is to be fully filled in from Sheet 11; the Plan of the End Motion is to be filled in from Sheet 41 and details. The plan of the Saddle is to be fully filled in from Sheet 29 and details; and the Loose Head is to be fully shewn, with clamp plate, from Sheet 21 and details. The Gantry is to be fully filled in from Sheet 45, and the Leading Screw and Back Shaft from Sheets 41, 44, 46, 47, 48 and 49.

The Gap Piece, near the Fast Head, can be removed, to give room for any large diameter short article to be mounted on a large face place, which is similar to the catch plate but larger in diameter. Boring and surfacing are usually done with the Gap open.

This Drawing should contain all that is necessary for taking off details for manufacturing the Lathe. A Sheet containing a Back view of the Lathe would be a good further exercise, shewing such things as cannot be shewn in the three views of this Drawing without confusion, and the remainder of the parts in outline. It is not however necessary in view of the full details shewn in the group drawings.

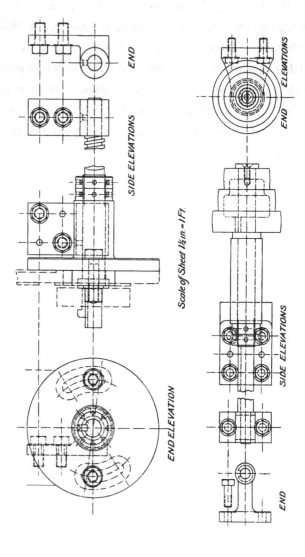

SHEET 44

8" LATHE, BRACKETS for LEADING SCREW and BACK SHAFT

END

SIDE ELEVATIONS

END ELEVATION

Scale of Sheet 1½ in = 1 Ft.

END ELEVATIONS

SIDE ELEVATIONS

END

SHEET 44 contains views of the brackets for the Leading Screw and the screw ends and fittings, and also the same features of the Back Shaft.

The Book scale is $1\frac{1}{2}''$ to 1 foot and the Drawing is to be made half-size on a double elephant sheet of paper which is also to contain the figures on Sheets 42, 47 and 49.

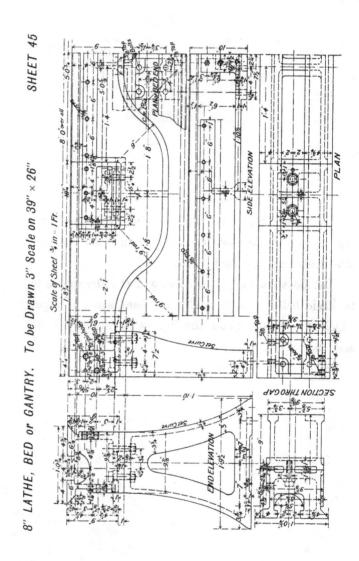

8" LATHE, BED or GANTRY. To be Drawn 3" Scale on 39" × 26"

SHEET 45

Scale of Sheet ¾ in = 1 Ft.

SHEET 45 contains views and sections for a full drawing of the Gantry, with Gap Piece, and Standards, and all fastenings for these; as also all drilling and other work, dimensioned finished sizes.

This drawing is to be made 3″ scale, and will fill a double elephant sheet. The figures on Sheets 46 and 48 are to be shewn mounted in position, and also separately on the same sheet. This Sheet would form a working drawing for the Pattern and Machine Shops, for the Gantry and its fittings, ready for the moving parts.

The Book figure is to $\frac{3}{4}$ inch scale, and the views are cut up to get them into the space of the page, but the drawing is to be made after the arrangement of Sheet 43; with the separate bracket figures in the spare space; the Gantry drawing views being placed as closely as convenient for lettering and dimensioning; and the bracket items, so that the print will cut up, for separate handling.

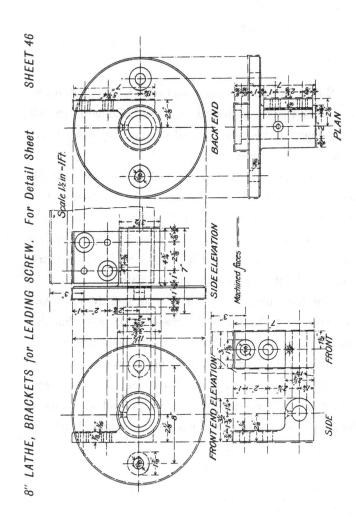

8″ LATHE, BRACKETS for LEADING SCREW.   For Detail Sheet   SHEET 46

SHEET 46 contains the Leading Screw brackets, the Fast Head end bracket in four views and the small end bracket in two views. These as before stated are to go into the Gantry sheet, both fixed in position and separately.

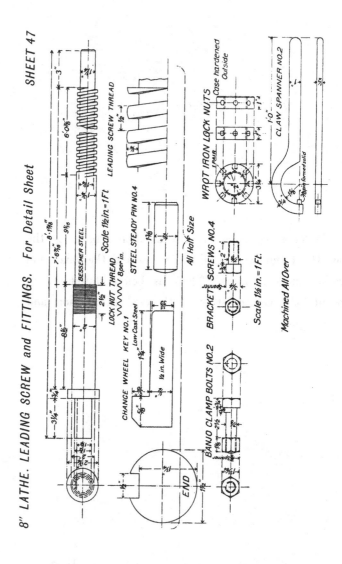

8" LATHE. LEADING SCREW and FITTINGS. For Detail Sheet

SHEET 47

SHEET 47 contains details of the Leading Screw, key for the change wheels, steady pins for the Screw bracket, Banjo clamp bolts, screws for the brackets, lock nuts and claw spanners for the same. Also sample threads of the Screw, etc.

The Book figures are $1\frac{1}{2}''$ scale, with the exceptions of the Screw end and key-way, the key, the steady pin, and the sample threads, all of which are half size and are to be drawn full size on the Drawing before mentioned which is to contain Sheets 42, 44, 47 and 49. The other figures on this sheet are to be drawn 3" scale.

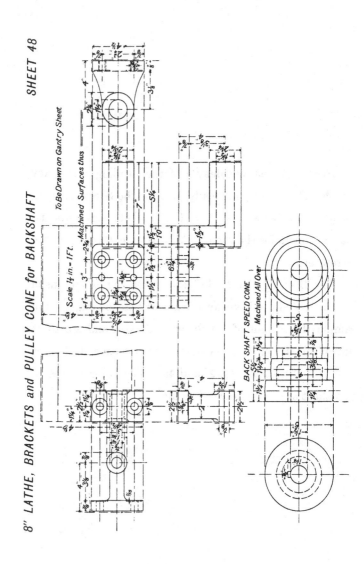

SHEET 48

8" LATHE, BRACKETS and PULLEY CONE for BACKSHAFT

To Be Drawn on Gantry Sheet

Machined Surfaces thus

Scale 1½ in. = 1 Ft.

BACK SHAFT SPEED CONE

Machined All Over

SHEET 48 contains the brackets and cone pulley for the Back Shaft. These figures are to be drawn on the Gantry Sheet, both in position and separately, as mentioned for Sheet 45.

They are to be drawn like the Leading Screw brackets below the Gantry figures, 3" scale, the same as the Gantry and so that they can be cut off the print.

The Book figures are $1\frac{1}{2}''$ scale.

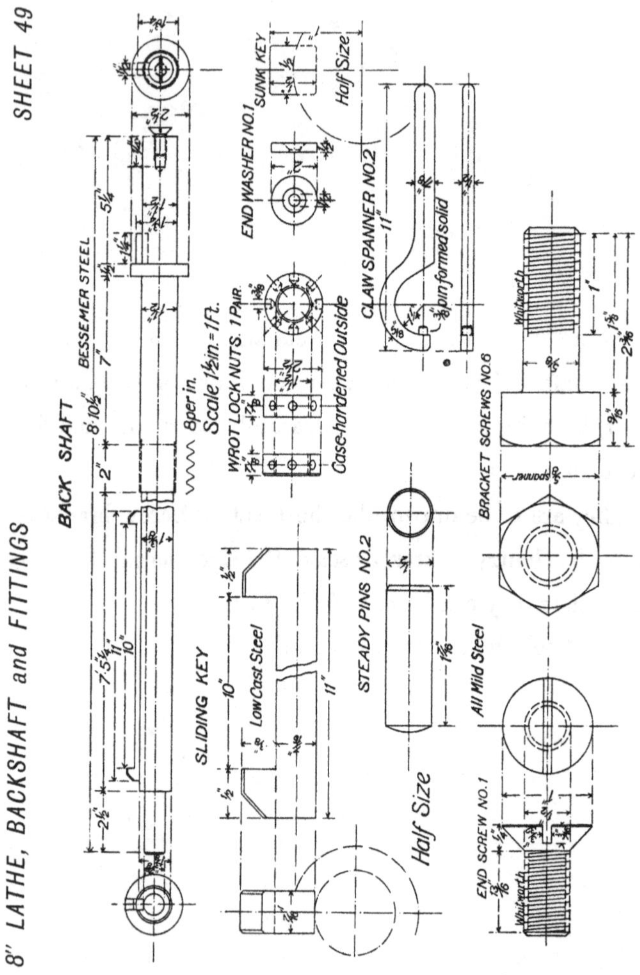
8" LATHE, BACKSHAFT and FITTINGS
SHEET 49
BACK SHAFT
BESSEMER STEEL
8 per in.
Scale 1½ in. = 1 Ft.
SLIDING KEY
Low Cast Steel
WROT LOCK NUTS. 1 PAIR
Case-hardened Outside
END WASHER NO.1
SUNK KEY
Half Size
CLAW SPANNER NO.2
⅜" pin formed solid
STEADY PINS NO.2
Half Size
BRACKET SCREWS NO.6
All Mild Steel
END SCREW NO.1
Whitworth
spanner

SHEET 49 contains details of the Back Shaft, lock nuts, end washer and claw spanners, to $1\frac{1}{2}''$ scale. It also contains the sliding key, steady pins for brackets, and screws for brackets, half size.

The drawings are to be made on the Detail Sheet of shafts, screws, etc., from Sheets 42, 44, 47 and 49, the scales to be 3″ and full size respectively.

8″ LATHE, LIST of CHANGE WHEELS
All $\frac{5}{8}$″ Pitch.   SHEET 50

	No. of Wheels	No. of Teeth	Outside Diameter	Depth of Teeth	2″ hole	1½″ hole	2″ wide	1½″ wide	Wrot iron	Cast iron
			″	″						
1	1	20	$4\frac{3}{8}$	$1\frac{13}{32}$	—		—		--	
2	2	25	$5\frac{3}{4}$	″	—		—		—	
3	1	25	$5\frac{5}{8}$	″		—		—		—
4	1	27	$5\frac{3}{4}$	″		—		—		—
5	1	28	$5\frac{15}{16}$	″	—		--			—
6	1	30	$6\frac{3}{8}$	″	—		--			—
7	2	30	$6\frac{3}{8}$	″		—		—		—
8	1	32	$6\frac{3}{4}$	″		—		—		—
9	1	34	$7\frac{1}{8}$	″		—		—		—
10	1	35	$7\frac{3}{8}$	″	—		—			—
11	1	35	$7\frac{7}{8}$	″		—		—		—
12	1	36	$7\frac{9}{16}$	″		—		—		—
13	2	40	$8\frac{3}{4}$	″		—		—		—
14	1	42	$8\frac{3}{4}$	″		—		—		—
15	1	45	$9\frac{3}{4}$	″		—		—		—
16	1	46	$9\frac{17}{32}$	″		—		—		—
17	1	48	$9\frac{15}{16}$	″		—		—		—
18	2	50	$10\frac{3}{8}$	″		—		—		—
19	1	54	$11\frac{1}{8}$	″		—		—		—
20	1	55	$11\frac{5}{16}$	″		—		—		—
21	1	57	$11\frac{3}{4}$	″		—		—		—
22	2	60	$12\frac{5}{8}$	″		—		—		—
23	1	63	$12\frac{29}{32}$	″		—		—		—
24	1	65	$13\frac{9}{32}$	″		—		—		—
25	1	70	$14\frac{9}{32}$	″		—		—		—

SHEET 50 is a Table of Change Wheels for Screw Cutting, from 28 threads per inch to 2″ pitch of thread. The teeth of the wheels are extra strong for the size of Lathe, being ⅝ inch pitch, and such as are suitable for heavy worm cutting; and the "counts" of the teeth are such as keep the diameters of the largest wheels within reasonable dimensions, considering the size of the lathe. There are 30 wheels in all which suit the combinations shewn on Sheet 51 for "eighth" screws, that is, multiples of eighths of an inch; and also the combinations shewn on Sheet 52 for all the Whitworth standard threads, including gas threads.

8″ LATHE, CHANGE WHEEL NUMBERS for WORM and "EIGHTH" SCREWS. Leading Screw ½″ pitch
SHEET 51

Multiplying and Reducing Nos. in heavy type. Motion wheels light.

	Pitch of Screw in inches	Multiplication of Gear to—	Reduction of Gear to—	No. on Lathe-spindle	No. on Stud	Multiplying and Reducing combination Nos.		No. on Leading Screw
1	2	4		30	40	$\begin{Bmatrix}30\\25\end{Bmatrix}$	$\begin{Bmatrix}60\\50\end{Bmatrix}$	30
2	$1\frac{3}{4}$	$3\frac{1}{2}$		30	40	$\begin{Bmatrix}30\\20\end{Bmatrix}$	$\begin{Bmatrix}60\\35\end{Bmatrix}$	30
3	$1\frac{1}{2}$	3		30	40	$\begin{Bmatrix}30\\20\end{Bmatrix}$	$\begin{Bmatrix}60\\30\end{Bmatrix}$	30
4	$1\frac{1}{4}$	$2\frac{1}{2}$		30	40	$\begin{Bmatrix}30\\40\end{Bmatrix}$	$\begin{Bmatrix}60\\50\end{Bmatrix}$	30
5	$1\frac{1}{8}$	$2\frac{1}{4}$		30	40	$\begin{Bmatrix}30\\40\end{Bmatrix}$	$\begin{Bmatrix}60\\45\end{Bmatrix}$	30
6	1	2		35	40	**30**	$\begin{Bmatrix}60\\40\end{Bmatrix}$	35
7	$\frac{7}{8}$	$1\frac{3}{4}$		**28**	40	**30**	**60**	**32**
8	$\frac{3}{4}$	$1\frac{1}{2}$		30	40	**40**	**60**	30
9	$\frac{5}{8}$	$1\frac{1}{4}$		30	40	**40**	**50**	30
10	$\frac{1}{2}$	1		30	40		60	30
11	$\frac{7}{16}$		$\frac{7}{8}$	**28**	40		60	**32**
12	$\frac{3}{8}$		$\frac{3}{4}$	**30**	40		60	**40**
13	$\frac{5}{16}$		$\frac{5}{8}$	30	40		60	**48**
14	$\frac{1}{4}$		$\frac{1}{2}$	30	40		40	**60**
15	$\frac{3}{16}$		$\frac{3}{8}$	30	40	**60**	**45**	**60**
16	$\frac{5}{32}$		$\frac{5}{16}$	30	50	**40**	**25**	**60**
17	$\frac{1}{8}$		$\frac{1}{4}$	30	40	**50**	**25**	**60**
18	$\frac{3}{32}$		$\frac{3}{16}$	30		$\begin{Bmatrix}40\\60\end{Bmatrix}$	$\begin{Bmatrix}30\\30\end{Bmatrix}$	**60**
19	$\frac{1}{16}$		$\frac{1}{8}$	30		$\begin{Bmatrix}50\\60\end{Bmatrix}$	$\begin{Bmatrix}25\\30\end{Bmatrix}$	**60**

SHEET 51 is a Table of combinations of change wheels for producing screw threads from $\frac{1}{16}$ in. pitch, advancing by $\frac{1}{32}''$ to $\frac{3}{16}''$ pitch, by $\frac{1}{16}''$ to $\frac{1}{2}''$ pitch, by $\frac{1}{8}''$ to $1\frac{1}{4}''$ pitch and by $\frac{1}{4}''$ to $2''$ pitch. These are termed "eighth" screws. Others are termed "odd" numbers.

8″ LATHE, CHANGE WHEEL NUMBERS for WHITWORTH SCREWS (Including Gas). Leading Screw $\frac{1}{2}$″ pitch

SHEET 52

Multiplying and Reducing Nos. in heavy type. Motion wheels light.

No. of Threads per inch	Multiplication of Gear to—	Reduction of Gear to—	Nos. of Gear on Lathe spindle	Nos. on Stud Idle wheels	Multiplying and Reducing combination Nos.		Nos. on Leading Screw
28		$\frac{1}{14}$	20		{60 / 63}	{30 / 27}	60
24		$\frac{1}{12}$	20		{60 / 50}	{30 / 25}	60
20		$\frac{1}{10}$	20		{60 / 50}	{30 / 25}	50
19		$\frac{1}{9.5}$	20		{60 / 57}	{30 / 36}	60
18		$\frac{1}{9}$	20		{60 / 54}	{30 / 36}	60
16		$\frac{1}{8}$	20		{48 / 60}	{36 / 30}	60
14		$\frac{1}{7}$	20		70	30	60
12		$\frac{1}{6}$	20		60	30	60
11		$1/5\frac{1}{2}$	20		55	30	60
10		$\frac{1}{5}$	20		50	30	60
9		$1/4\frac{1}{2}$	20		45	30	60
8		$\frac{1}{4}$	20		40	30	60
7		$1/3\frac{1}{2}$	20		35	30	60
6		$\frac{1}{3}$	20		Idle	wheels	60
5		$1/2\frac{1}{2}$	20		,,	,,	50
$4\frac{1}{2}$		$1/2\frac{1}{4}$	20		,,	,,	45
4		$\frac{1}{2}$	20		,,	,,	40
$3\frac{1}{2}$		$1/1\frac{3}{4}$	20		,,	,,	35
$3\frac{1}{4}$		$1/1\frac{5}{8}$	20		,,	,,	65
3		$1/1\frac{1}{2}$	20		,,	,,	30
$2\frac{7}{8}$		$1/1\frac{7}{16}$	30		46	32	30
$2\frac{3}{4}$		$1/1\frac{3}{8}$	30		55	40	30
$2\frac{5}{8}$		$1/1\frac{5}{16}$	30		42	32	30
$2\frac{1}{2}$		$1/1\frac{1}{4}$	30		50	40	30
$2\frac{1}{4}$		$1/1\frac{1}{8}$	30		45	40	30
$2\frac{1}{8}$		$1/1\frac{1}{16}$	30		34	32	30

SHEET 52 is a Table of combinations of change wheels for cutting all the Whitworth standard threads, including gas threads.

8" LATHE, CHANGE WHEELS. All cut Teeth.   Scale $1\frac{1}{2}$" = 1 Ft.   SHEET 53

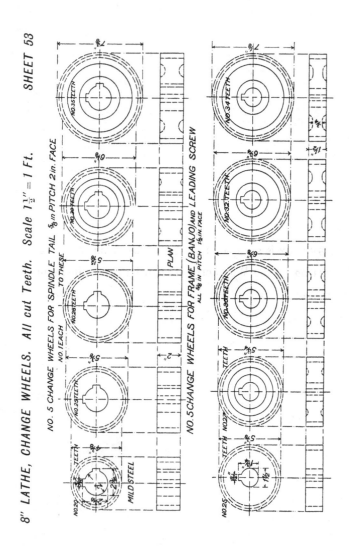

NO. 5 CHANGE WHEELS FOR SPINDLE TAIL $\frac{5}{8}$ in PITCH 2 in. FACE

NO. 5 CHANGE WHEELS FOR FRAME (BANJO) AND LEADING SCREW
ALL $\frac{5}{8}$ in PITCH $1\frac{1}{2}$ in. FACE

SHEET 53 contains five change wheels for the Lathe Spindle Tail fitting, and five sizes of Banjo and Leading Screw change wheels from $7\frac{1}{8}''$ to $7\frac{9}{16}''$ outside diameter. The Book figures are to $1\frac{1}{2}''$ scale, but double elephant sheets of drawings are to be made of the wheels on this sheet and Sheet 55, all full size, with five teeth drawn in each figure to Sheet 54 which is half size in the Book figure. The radii of the " points" are the thickness of the tooth on the pitch line and the centres of the radii are on the pitch line. The radii of the roots are one and a half the half pitch and the centres are the centres of the next teeth, on the pitch line. The teeth are to be drawn for the purpose of shewing the shape of the "spaces" and the cutters.

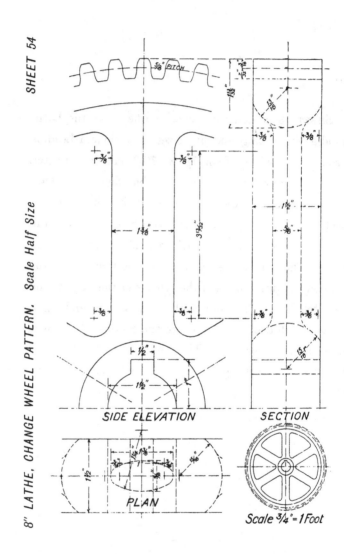

SHEET 54 contains a specimen boss, arm, rim, and teeth, for the change wheels which have arms, either four or six; and also the hole and key-way for the key-bosses on which they are mounted when working.

8″ LATHE, CHANGE WHEELS for BANJO and LEADING SCREW   SHEET 55

ALL 5/8 IN. PITCH 1½ IN. FACE

SOLID DISC. PATTERN

No. Wheels	Diam.	No. Teeth
1	7 5/8	35
1	7 15/16	36
2	8 3/8	40
1	8 3/4	42

FOUR ARM PATTERN

No. Wheels	Diam.	No. Teeth
1	9 3/8	45
1	9 9/32	46
1	9 5/8	48
2	10 3/8	50

SIX ARM PATTERN

No. Wheels	Diam.	No. Teeth	No. Wheels	Diam.	No. Teeth
1	11 1/8	54	1	12 25/32	63
1	11 9/16	55	1	13 3/16	65
1	11 3/4	57	1	14 9/32	70
2	12 3/8	60			

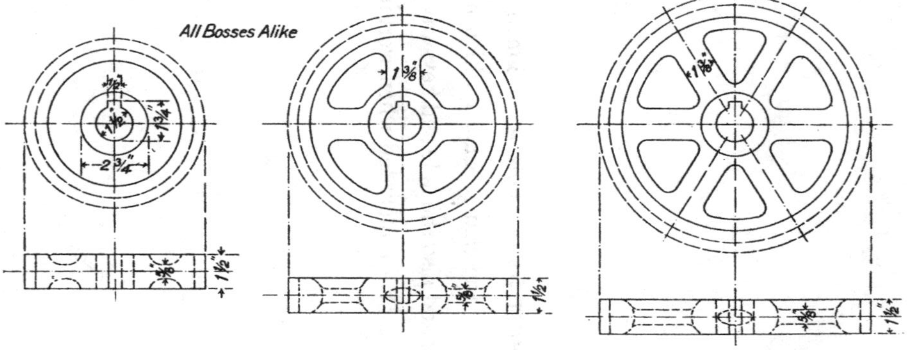

All Bosses Alike

SHEET 55 contains one solid change wheel, one with four arms and one with six arms. The outside dimensions are given in figures and the depths and radii of the teeth are shewn in the specimen figure on Sheet 54.

All these wheels are to be drawn full size on a double elephant sheet and are to have five teeth drawn into each rim to shew the shape and sizes of the spaces for the cutters, the whole of the wheel teeth being machine cut. The holes and key-ways are to suit the key-bushes in which they are mounted in working.

8INCH LATHE. DETAILS of CAST IRON WORK. To Be Drawn Half Size as SHEET 9.   SHEET 17.
And on Detail Sheets

8INCH LATHE. DETAILS of CAST IRON WORK. (To Be Drawn Half Size as SHEET 9.) SHEET 16.
(And on Detail SHEET.)

8INCH LATHE. DETAILS of FAST HEAD FINISHED SIZES   SMITHS WORK   SHEET 19.

8INCH LATHE. DETAILS of FAST HEAD FINISHED SIZES   BRASS WORK

8INCH LATHE. ARRANGEMENT of DRAWING of FRONT of SADDLE MOTIONS (ON 39 IN. x 26 IN.) SHEET 26

8INCH LATHE. ARRANGEMENT of DRAWING of BACK of SADDLE MOTIONS (ON 39 IN. x 26 IN.) SHEET 30.

8INCH LATHE. BACK of SADDLE MOTIONS   (To Be Drawn Half Size To Sheet 30) SHEET 31.

8INCH LATHE. LOOSE HEAD DETAILS of STEEL WORK. Scale 1½ in. = 1 Ft.   SHEET 25.
To Be Drawn Full Size on Detail Sheet, and for Detail in Sheet 20.

8INCH LATHE. DETAILS of SADDLE FITTINGS. (For Sheets 26 & 30 and Detail Sheets.)   SHEET 34.
DOVETAIL SLIP CAST IRON.

8INCH LATHE. DETAILS of SADDLE MOTIONS. (For Sheets 26 & 30 and Detail Sheet.)   SHEET 35.